"Self-driving cars are coming quickly, so we need to control them somehow. To reduce their dangers, Paulo and Kirchmair deftly weave together contemporary moral psychology, the AI technology behind self-driving cars, and concrete proposals for policy and law. This book is a multi-disciplinary tour-de-force."

Walter Sinnott-Armstrong, *Duke University, USA*

"Do we really need self-driving cars, and how should they behave in risky situations? This clearly written and very engaging book tackles these intriguing questions head on. Norbert Paulo and Lando Kirchmair offer the reader an exciting whirlwind tour through the history of car crashes and transport us into the future of the ethics of autonomous driving. Along the way, both moral and legal issues are explored. Paulo and Kirchmair also investigate fascinating methodological questions about how we should approach the ethics of autonomous driving. At the end of the book, they offer clear-headed recommendations for policy makers. Given the widespread interest in the ethics of self-driving cars, it is somewhat surprising that while an enormous number of articles have been published on the subject, not many books have been published so far. Of the few books available on this important topic, this may very well be the best so far."

Sven Nyholm, *LMU Munich, Germany*

"This valuable book provides an insight into the ethics of self-driving cars that is accessible and timely – and a must-read for those charged with regulating self-driving cars."

Julian Savulescu, *National University of Singapore*

"This book is a timely contribution to the debate on the ethics of autonomous vehicles focusing on the importance of trolley-like scenarios. Well-written and well-argued, with practical recommendations."

David Černý, *Czech Academy of Sciences, Czechia*

MORAL DILEMMAS INVOLVING SELF-DRIVING CARS

Experts believe that self-driving cars will be much safer than conventional cars. However, there will still be some accidents, even fatal ones. The innovation of self-driving cars gives us the new, amazing, and, at the same time, arduous opportunity to decide in advance what should happen in which accident situation. It will force us to decide who must die in order to save others in dilemma situations.

This book prepares readers to take an informed stance on the difficult moral dilemmas involving self-driving cars. It will convince them that the ethical and legal challenges posed by self-driving cars are real, novel, and unavoidable—and that they require a solution. It presents the problems involved in programming self-driving cars for dilemma situations, putting the reader in the position to develop their own well-informed opinion on the issues. The book also discusses the astonishing discrepancies between existing regulations for collisions involving self-driving cars in some jurisdictions and empirical findings about what people think about how they should be regulated. Finally, it presents advice for policymakers around the globe on how to regulate moral dilemmas involving self-driving cars.

Moral Dilemmas Involving Self-Driving Cars will be of interest to a broad range of students, scholars, and policymakers interested in the ethics of self-driving vehicles.

Norbert Paulo is DFG Heisenberg Fellow at the Department of Philosophy at Ludwig-Maximilians-University Munich. He obtained his PhD at the University of Hamburg. His PhD dissertation has been published as *The Confluence of Philosophy and Law in Applied Ethics* (2016).

Lando Kirchmair is Deputy Professor for National and International Public Law with a Focus on the Protection of Cultural Heritage at the University of the Bundeswehr Munich. His habilitation thesis—*Rethinking the Relationship between International, EU and National Law: Consent-Based Monism*—was published in 2024.

MORAL DILEMMAS INVOLVING SELF-DRIVING CARS

How to Regulate Them and Why Your Opinion Matters

Norbert Paulo and Lando Kirchmair

Routledge
Taylor & Francis Group
NEW YORK AND LONDON

Cover image designed by Richard Yabes Soedarno

First published 2025
by Routledge
605 Third Avenue, New York, NY 10158

and by Routledge
4 Park Square, Milton Park, Abingdon, Oxon, OX14 4RN

Routledge is an imprint of the Taylor & Francis Group, an informa business

ISBN: 978-1-032-77881-5 (hbk)
ISBN: 978-1-032-77882-2 (pbk)
ISBN: 978-1-003-48521-6 (ebk)

DOI: 10.4324/9781003485216

Typeset in Sabon
by Newgen Publishing UK

CONTENTS

ACKNOWLEDGMENTS

We are very grateful to the Digitalization and Technology Research Center of the Bundeswehr (dtec.bw) funded by the European Union—NextGenerationEU. It may come as a surprise that we have mentioned the source of funding first. However, without the grant for our research project EMERGENCY-VRD, this book would not exist. Dtec.bw and the University of the Bundeswehr Munich provide an exceptional research environment—intellectually stimulating and open, truly interdisciplinary, and with incredible technological possibilities.

In this research project, we—a philosopher and a legal scholar—are working together with Matthias Gerdts—a mathematical engineer—to investigate the ethics of self-driving cars. The project naturally led to the writing of this book, in which we present the main arguments for deciding how to program self-driving cars in moral dilemmas. We are deeply indebted to Matthias Gerdts. We could not imagine a better partner to bridge the gap between the humanities and the sciences in the field of self-driving cars.

It was extremely important for the development of the manuscript that we thought through every single idea together before we wrote it down, of course only to rewrite it again and again. In the end, we can no longer say who wrote what. And so we contributed equally to the book from beginning to end.

We would like to mention the participants in our project workshops for their support as well as those who commented on our work at other academic events. Special thanks go to Iris Eisenberger, Paul Schiff Berman, Yochanan Bigman, Laura A. Dickinson, Anne Meuwese, Leonie Möck, Sven Nyholm, Leon Sütfeld, and Matthias Uhl. We also benefited greatly from very helpful and detailed comments on earlier drafts by Walter Sinnott-Armstrong, Thomas Pölzler, and Sebastian Krempelmeier. Finally, Helen Heaney was enormously helpful in avoiding many errors in the English language and beyond. We would like to emphasize that these are the people who helped to improve the manuscript. Moreover, we warmly thank Richard Yabes Soedarno for designing the cover image and Nina Augustin for compiling the index.

Last but not least, a huge inspiration for writing this book has been you: every single reader we imagined while writing the book, readers who are interested in moral dilemmas and self-driving cars, and readers who are ready to make up their minds about difficult moral issues that will soon be part of our public life. So, thank you.

PROLOGUE

How should autonomous vehicles be programmed for dilemma situations? Should they kill an elderly person if this would save a child, or one person if this would save two? This is neither a horror story nor science fiction. The subject matter of this book will, very soon, directly affect you, us, and everyone else. In the end, we the people will have to make this decision, and the aim of this book is to prepare you for making it, not spontaneously or when flustered, but after having considered the relevant arguments. We want you to be able to make a thoroughly informed decision about some of the hardest moral questions that societies will face worldwide in the coming years, namely about dilemmas involving autonomous vehicles.

Experts believe that self-driving cars will be much safer than conventional cars: they will cause far fewer accidents involving far fewer fatalities. However, some accidents will still occur, even fatal ones. And some of these will represent dilemmas in which all available options lead to comparable harm. Just imagine an autonomous vehicle with defective brakes. Because it cannot brake, it will either run over a pedestrian, which would probably kill them, or crash into another car, which would probably kill one passenger in that car. There are no easy solutions to such moral dilemmas. When human drivers are in such a situation, they typically react spontaneously and emotionally; the time is too short for a controlled reaction, let alone for a thought-through moral decision. For the first time in human history, the rapidly advancing technology behind self-driving cars is giving us the opportunity to decide what to do in such dilemma situations in advance. We will soon be able to program cars to do what we regard as the best moral solution to such dilemmas.

This is a great chance. But it is also a great responsibility. Societies worldwide are now burdened with decisions relating to how to program autonomous vehicles for dilemma situations in something as mundane as everyday traffic. Like it or not, we the people have to decide who to save and who to kill.

We do not promise that you will find it easy to take a stance on this problem once you have read this book. But we do promise that you will understand why it is impossible for society not to make the decision, and why ethicists, lawyers, and—most importantly—lawmakers need to know what people think about moral dilemmas involving self-driving cars.

1

THE POSSIBILITY AND NECESSITY OF REGULATION

Preparing Ourselves for Urgent Moral Dilemmas

1.1 A Short History of Car Accidents

It was in 1885 when German engineer Carl Benz presented his 'Benz Patent-Motorwagen Nummer 1.' In this first 'Benz,' a combustion engine replaced draught animals as the means of propulsion. It is widely regarded as the first practicable 'automobile' and was the beginning of what is known today as 'Mercedes-Benz.' However, it took three more years and the courage of the engineer's wife to show the world exactly how practicable automobiles are. On 5 August 1888, Bertha Benz wanted to visit her mother in Pforzheim. Because this was about 60 miles away from their home, she decided to use 'Motorwagen Nummer 3.' Without the knowledge of her all too cautious husband—and without a driving license—she and their two sons embarked on the first ever long-distance trip by car (Deutsches Patent- und Markenamt 2022). The invention of the automobile changed individual and mass transportation alike, ultimately facilitating completely new ways of life. For example, the rapid development of suburbia would have been impossible without cars and buses with combustion engines.

It was soon realized, though, that a growing number of automobiles led to more accidents. Draught animals do not usually run people over. They have a natural tendency to avoid running into someone or something, even if the driver does not realize that the vehicle is heading toward an obstacle. Legend has it that well-trained draught animals even brought their drunken drivers home safely. Because cars do not have such natural instincts or talents, they are more likely to cause accidents when their human drivers make mistakes.

On 'a balmy summer afternoon' on 17 August 1896, Bridget Driscoll, the 44-year-old 'wife of a labourer, of Old Town, Croydon' (*The Guardian* 2014), became the first victim of an automobile accident (McFarlane 2010; Porter 1998). Mrs. Driscoll was spending a day out at Crystal Palace in south-east London when she was hit by a 'Roger-Benz.' This car was based on Carl Benz's invention and manufactured by the Frenchman, Émile Roger. It was part of an exciting motoring exhibition at Crystal Palace. The driver was not drunk, but perhaps 'speeding.' A witness to the accident stated that the car went at a 'tremendous pace,' at least 'as fast as a good horse could gallop' (McFarlane 2010). The speed of the car

DOI: 10.4324/9781003485216-1

was either the allowed 4 mph or, allegedly, more than twice as fast due to a 'mischievously removed' speed belt (Porter 1998). In any case, the car was going too fast for the victim, who could not be saved although the car bell had been rung and the driver shouted 'stand back!'. The role of the driver was investigated in the trial following the accident. The jury 'returned a verdict of accidental death,' however, as conflicting reports about the speed of the car could not be resolved (*The Guardian* 2014).[1]

Of course, this first fatal car accident was not the last. On 12 February 1898, the Brighton businessman Henry Lindfield, a sales representative for International Cars, became the first driver to be killed in a car accident. Driving at up to 17 mph, Mr. Lindfield lost control of his vehicle and crashed into a tree. Apparently, it was only the second or third time that he had driven a car. Mr. Lindfield died a few hours after the crash in hospital due to shock after having his leg amputated. His teenage son, who had gone along for the ride, only had minor injuries (Information Britain 2010).

Lindfield's accident involved an electrically powered car. Hence, until 1899 the argument could be made that petrol-powered cars were safer. If someone had placed their hopes in that premise, they were soon to be disappointed. A plaque reminds us of an accident that took place on 25 February 1899 in Grove Hill, London. The driver Edwin Root Sewell and a passenger named Major Ritche were in a car accident caused by a broken front wheel. The driver died instantly; Ritche died later from his injuries; the other four passengers only had minor injuries. Sewell and Ritche were the first persons to die in a petrol-powered motor vehicle accident (London Remembers 2024).

But not only British people succumbed to accidents with cars. On 14 September 1899, Henry Hale Bliss, a 69-year-old New York real estate man, became the first (known) victim of the automobile revolution in the United States when he was getting off a tram near Central Park in downtown New York. Mr. Bliss, a gentleman *par excellence*, stepped out of the tram, turned, and offered a hand to his companion, Mrs. *Lee*. This kindness was fatal. An electrically powered cab, which made relatively little noise, hit and ran him over. Even though the passenger in the cab was David Orr Edson, a doctor (and son of a former mayor of New York), who immediately came to his aid, Mr. Bliss died the next day in Roosevelt Hospital.[2]

1.2 Auto-Autos Are (a Bit) Like Carriages

More than 125 years after the first fatal car accident, modern road traffic causes approx. 1.35 million fatalities per year worldwide (World Health Organization 2018). Many of these—approx. 90%—are due to mistakes made by the drivers, for example, because they were speeding, violated priority rules, or because they were so tired or intoxicated that they could not react quickly enough (Dingus et al. 2016).[3] There is a technical innovation that promises to overcome these human errors and limits by making cars as safe as carriages again, namely the development of automated vehicles. The 'auto-auto,' or 'autonomous automobile,' as a funny German nickname[4] goes, is one designation for an autonomous, i.e., self-driving, car that is both propelled *and* controlled 'automatically,' hence, without human intervention. In other words, they combine and automate the abilities of the driver and the draught animal. Horses and humans have now been fully substituted by technology. Automated vehicles do not need human drivers anymore, and when there are no human drivers, they cannot make mistakes while driving. The automated control technology thus makes car rides safer by making cars more like carriages with draught animals again (in one respect, at least) (Papadoulis, Quddus, and Imprialou 2019).

Combining the luxury of relaxing while taking a train or plane with the individual flexibility of going by car is tempting. By removing the burden of having to constantly pay attention while driving, nothing less than another mobility revolution is about to unfold with the rise of self-driving vehicles. Such vehicles allow for more inclusion as driver's licenses and driving abilities become relics of the past. Everyone will be able to go by car, children, and the elderly alike, without being dependent on someone driving them.[5]

1.3 How Dangerous Are Self-Driving Cars?

Can self-driving cars really help make road traffic safer? Are there not many accidents involving such cars? The first widely reported fatal accident with a Tesla using its self-driving technology, the so-called 'Autopilot' mode, occurred in May 2016 (Bonnefon 2021, chap. 17). A large truck had crossed the highway in front of the Tesla at an intersection. The Tesla did not brake and continued driving under the truck, losing its roof and running off the road until it hit a pole. As an investigation of the accident revealed, the fatally injured driver was not monitoring the car properly. Although the car had alerted the driver several times to keep his hands on the wheel, he only did so for a total of 25 seconds in the 37 minutes before the accident. This illustrates that Tesla's Autopilot can operate a car pretty much on its own. However, it is important to note that its capacity is very limited. It is designed for relatively clear and simple circumstances. In this case, the Autopilot was designed for limited-access highways only; it was not designed for intersections. The report found that the Autopilot was not malfunctioning but that it was not meant to be driving the car in such circumstances. Instead, the blame was passed onto the fatally injured driver.

Or take the first accident in which a pedestrian fell victim to an automated car. In 2018, a car belonging to the Uber fleet was driving in automated mode but with an emergency backup driver behind the wheel whose job it was to supervise the car. The car was driving down a four-lane road in Tempe, Arizona, at 40 mph when it struck and killed Elaine Herzberg (Levin and Carrie 2018). Since it was dark, and the victim was trying to cross the road while pushing her bicycle, the problem might have been that the backup driver could not react in time because she could not see Mrs. Herzberg in the dark. But what we do know now, after the police released two videos (Levin and Carrie 2018), is this. The car's interior camera filmed the operator during the 15 seconds leading up to the accident, and the footage seems to show that she was not looking at the road. As investigations later revealed, an episode of *The Voice* was being streamed on the operator's phone when the accident happened. However, it remains unclear whether she was watching or merely listening to the episode (Somerville and Shepardson 2018). Apparently, she had placed so much trust in the car that she was not paying enough attention to what was happening on the road. But why did the car not brake, or at least alert the driver? The car did realize that there was an obstacle ahead six seconds before the crash and wanted to perform an emergency stop. But it did not brake. Why not? Because its settings were such that only the backup driver was authorized to initiate an emergency stop and the car only alerted the operator 0.2 seconds before impact. After that tragic accident, Uber changed the settings. Now the cars are supervised by two backup drivers (again) and can perform emergency stops (Bond 2018). The backup safety driver was charged with negligent homicide in 2020 after pleading guilty to reckless endangerment, a lesser charge. Interestingly, prosecutors had already said in March 2019 that Uber would not face criminal charges (Shepardson 2023).

To be able to evaluate these and similar accidents, of which there certainly will be many more in the future, it is important to understand that there are different levels of automation. These levels go hand in hand with the different roles people have in the driving process. According to the Society of Automotive Engineers, there are six levels (SAE International 2021). Level 0 means no driving automation. This would be the conventional automobile as invented by Carl Benz. The higher the level of automation, the more driving tasks the car is able to perform on its own. At level 1, the car has driving assistants such as adaptive cruise control or lane departure warning systems. Virtually every modern car has some such assistants. At level 2, the vehicle is able to do relatively simple things on its own, such as parking or overtaking maneuvers under favorable conditions, although the driver still has to monitor the car when it is operating automatically. Most accidents that are reported in the media these days happen with cars on level 2. A report by the National Highway Traffic Safety Administration shows that during a period of ten months in 2021 and 2022, fewer than 400 accidents with level 2 vehicles were reported to the U.S. authorities; at least six of these had fatal consequences (NHTSA 2022). Both the Tesla and Uber cars involved in the deadly accidents in 2016 and 2018 were on this level, as are almost all driving automation systems that are already on sale for the general public. Hence, the accidents you have read about in the news most likely happened because the technology was not yet ready for the situations the cars faced. Even though it was technological errors that caused these accidents, the human drivers had, in fact, failed to fulfill their responsibilities.

Introduced in early 2022, Mercedes-Benz's autopilot system—the so-called 'Drive Pilot'—is the first level 3 system that complies with the international UN R157 standard and is available to the general public, at least in Germany (Chan 2021) and since January 2023 also in Nevada. California soon followed suit and was indeed the state where the first Mercedes was sold that actually legally allowed drivers to put their hands anywhere but on the steering wheel (Jones 2024). The German luxury brand was only outpaced by Japan's Honda, which offered level 3 cars in Japan as early as 2021 (Honda 2021). At level 3, drivers do not have to constantly monitor the car when it is operating automatically, although they must be able to resume control whenever the car asks them to take over again, which implies that they must not be intoxicated, asleep, or distracted. Even though the Drive Pilot is a level 3 system, its use is still rather limited. So far, it has only been cleared by the authorities for use on certain '*Autobahnen*' in Germany and freeways in California and Nevada. It is only allowed to operate up to a speed of 60 km/h (and 40 mph), under favorable weather conditions and as long as there are no roadworks or other unusual circumstances. In other words, its use is pretty restricted. At least unlimited speeding, which is traditionally allowed on German highways will not be possible with the Drive Pilot. But it might be rather beneficial in traffic jams. Driver and passengers will be able to leave the unnerving stop-and-go to the Drive Pilot and watch *The Voice*, say. In addition, technology is a moving target and new improvements can be introduced more slowly or more quickly than expected. For example, Mercedes recently announced that the next step will be to go from the current 60 to 90 km/h (approx. 55 mph), and until the end of the 2020s, this shall be extended to 130 kmh (approx. 80 mph). Since December 2023, Mercedes has also been testing level 3 cars in Beijing (Prawitz 2023). The higher speed for the level 3 system should be available by the end of 2024. Not quite Lewis Hamilton style yet, but the technology is progressing.

At level 4, people in the car are no longer drivers but passengers. The cars can even drive without any passengers but their ability to drive automatically is still limited to certain environments or favorable conditions. At level 5, the car has full driving automation, meaning that it can perform the entire driving task completely alone and in all environments. At this final stage, it is not even possible for passengers in the car to intervene anymore. According to the National Highway Traffic Safety Administration, during the same ten-month period in 2021 and 2022, 130 accidents with cars on levels 3 to 5 were reported in the United States, none of which had fatal consequences. As the Drive Pilot and similar systems are not yet cleared for sale to the public in the United States, these 130 accidents happened during test drives and under human supervision (NHTSA 2022). Waymo, Google's sister company under the Alphabet umbrella, operates a level 4 'robo-cab' service in Phoenix, Arizona as well as San Francisco, and Los Angeles, California. China's Baidu robot taxi will take you to your destination within the Beijing Economic-Technological Development Area. Waymo recently claimed that their self-driving cars, with 7.1 million miles on the road, have already caused fewer accidents than human-driven cars. They reported an estimated 17 fewer injuries and 20 fewer police-reported crashes compared to human-driven cars. In other numbers, Waymo claims that considering these miles, their cars—with no human behind the steering wheel and with human passengers—had an 85% reduction in injury-causing crash rates and a 57% reduction in police-reported crash rates (Waymo 2023). While these numbers are certainly impressive, they somehow prove the point that self-driving cars are still involved in accidents.[6]

It is a matter of fierce debate on how quickly cars with full driving automation at level 5 will be available for the general public and taxi companies, etc. to purchase. While Tesla shocked other car manufacturers early on by claiming it would produce 'autonomous cars' sometime soon, its Autopilot is a rather typical level 2 assistance system. Nevertheless, optimistic experts consider that level 5 vehicles might become a commercial reality in some countries by the late 2020s (Litman 2020). Others hold that full driving automation is much more complex and that it will take much longer until it becomes a reality (Winton 2022). German lawmakers are quite convinced, however, that despite some skepticism, now is the time to enact comprehensive laws regulating self-driving cars (including the higher levels of automation). Therefore, it is high time to steer public debate in order to ensure that what the general public thinks about moral dilemmas involving self-driving cars is included in the process.[7]

For the purposes of this book, we assume self-driving vehicles to be at level 5. As level 5 systems do not yet exist, we are talking about the mid-term future. Even though some of the issues we will discuss can also arise with cars at lower levels of automation, we always assume that it is not a human driver but the automated car that does or does not brake or that does or does not swerve, etc.[8]

1.4 Ethical Reasons to Allow Self-Driving Vehicles

If the promise holds that self-driving vehicles will significantly reduce traffic accidents and, thus, fatalities, there are strong moral and legal reasons to allow them on the road. This is relevant not only for those of us who drive cars but also for those otherwise involved in road traffic, such as cyclists or pedestrians. It is, in fact, relevant for everyone.[9]

But the big question is whether self-driving vehicles will actually significantly reduce the number of accidents and fatalities. This is very difficult to answer, not least because it is hard to compare the statistics for accidents with self-driving cars with those for accidents with conventional cars. For example, self-driving cars are mostly tested under favorable conditions—in good weather, say. What is more, they are often tested on highways or other roads where accidents are relatively rare. Plus, the systems are almost always used in pretty new models that are equipped with state-of-the-art safety features. It is, thus, clear that we cannot simply compare the safety statistics for self-driving cars operated in test mode with the statistics for all conventional cars—old and new—driving under all weather conditions—good and bad—and on all types of roads—with high and low accident rates. So keep that in mind when large companies offer self-reported data about the safety of their cars. Despite the great progress, it always helps to remain critical. It might take years of gathering data about specific driving tasks before we can be certain that self-driving vehicles really are significantly safer than conventional cars in all circumstances (European Commission Expert Group 2020, chap. 1).

Despite that, we are all too well aware that people are very prone to error. We all behave incredibly stupidly from time to time. And it is such stupid mistakes that lead to a huge number of fatalities in car traffic. A central premise of self-driving vehicles is that traffic will be safer, simply because such vehicles do not speed, make phone calls, drive under the influence of alcohol or drugs, fall asleep at the wheel, or watch videos while driving. If this is true, there are strong moral reasons that speak for allowing them on the road.

The moral reasons for having self-driving cars are pretty straightforward. Take the old moral precept *primum non nocere* ('first, do no harm'), which dates back to the Hippocratic Oath at least (approx. fifth to third-century BC), and the *neminem laedere* ('harm no one') principle, first formulated by the Roman jurist Ulpian a few centuries later. Today, the same idea is captured in the principle of non-maleficence, which prohibits everyone from thwarting, defeating, or setting back the interests of others in medicine as well as in all areas of human conduct. As these traditional moral principles make clear, the prohibition of harm to others is one of the most important pillars of morality. It is also central to modern understandings of what the state can and should regulate.

In his 1859 book *On Liberty*, the English philosopher John Stuart Mill famously argued that the 'only purpose for which power can be rightfully exercised over any member of a civilized community, against his will, is to prevent harm to others' (Mill 1965, 223). This 'harm principle,' which stands behind modern understandings of individual rights and how they can be limited, also provides strong reasons for the transition from conventional cars to self-driving ones. Mill intended the principle to regulate relations between governments and individuals. But he also thought of it being applied prospectively to prevent people from causing harm to others. Often, we do not know with certainty what will cause harm and what will not. All that we know is that something *risks* causing harm. For Mill, the greater the harm that is risked, and the greater the probability that this risk materializes, the stronger the harm principle speaks against taking or allowing the risk (Brink 2007). Not every car journey kills people, of course, but every journey includes the risk of killing someone. As being killed is arguably one of the worst harms there is, the harm principle clearly demands the regulation of traffic. Similarly, the harm principle provides strong reasons for the adoption of self-driving cars, at least if they are less likely to cause fatal accidents than conventional cars and if the overall benefits of regulation exceed its costs.

Along these lines, the high-level German Ethics Commission on Automated and Connected Driving concluded in its 2017 report that '[t]he introduction of more highly automated driving systems, especially with the option of automated collision prevention, may be socially and ethically mandated if it can unlock existing potential for damage limitation' (Ethics Commission 2017, 11).

1.5 The Legal Case for Self-Driving Vehicles

From a legal perspective, the right to life is (one of) the most fundamental individual right(s). Everyone's right to life is enshrined in the Universal Declaration of Human Rights as well as in constitutions worldwide. Interestingly enough, Article 2 of the European Convention on Human Rights (ECHR) not only prohibits the state from intentionally ending human life but also obliges the state to take certain precautionary measures to prevent dangerous situations. The state, thus, has a duty to protect individuals from threats, and this includes risks imposed by other individuals. However, it is not easy to state for every concrete case whether this duty to protect applies. According to the European Court of Human Rights, it includes protective measures for individuals whose lives are threatened by environmental hazards as well as dangerous activities.[10]

The state's duty to protect is, of course, limited. For instance, the state does not have to prohibit road traffic simply because it is dangerous. A certain risk is regarded as being part of life. Yet the state must take legal measures to regulate the dangers emanating from road traffic.[11] Establishing appropriate and effective traffic regulations, such as a maximum blood alcohol level, is necessary to protect individuals from particular dangers, namely if there is a risk of serious and lethal accidents, which is known by the state. Such regulations can be seen as preconditions for the legality of road traffic. If they were not in place, road traffic would have to be forbidden. To the extent that self-driving vehicles fulfill their promise of significant increases in road safety, it can be assumed that the approval of such vehicles falls under the state's duty to protect life (Article 2 ECHR).[12]

Some even argue that if, or once, self-driving vehicles are safer than regular cars, human-operated cars should be forbidden. As Robert Sparrow and Mark Howard put it, once self-driving cars function as expected, human drivers would be 'like drunk robots' (Sparrow and Howard 2017). In other words, people should be banned from driving as they are a safety hazard in comparison with functioning self-driving cars. This is, of course, rather speculative. Many other considerations also play a role, such as the availability of self-driving cars for everybody who currently owns a car. Nevertheless, during her tenure as chancellor of Germany, Angela Merkel predicted as early as 2017 that 20 years from then, people would not be allowed to drive on public roads without a special permit that goes far beyond today's driver's license (Vitzthum 2017). Be that as it may, if self-driving vehicles promise to significantly reduce the number of fatal car accidents, there are strong moral *as well as* legal reasons to have them on the road.

1.6 A Caveat Before We Proceed: Algorithmic Bias, Privacy, and Other Tenacious Issues

These strong reasons for introducing self-driving vehicles on public roads are valid only with one important caveat. Self-driving cars must not introduce new risks that conventional cars do not. Have you heard of algorithmic bias? Timnit Gebru, Google's 'AI star ethicist,' announced on Twitter on 2 December 2020 that Google had forced her out of the company. More than 1,400 Google employees and other supporters actually signed a letter of protest on 4 December 2020. The letter expressed support for Dr. Gebru and stated that the termination of her position followed 'unprecedented research censorship' (Google Walkout For Real Change 2020). Why was Dr. Gebru so famous in the first place, and why is her research apparently so inconvenient for big tech companies? In 2018, she co-authored a very important paper on ethical research on artificial intelligence (AI). In the paper she published with Joy Buolamwini, another leading woman researching the ethics of AI, they showed that facial recognition software used by major tech companies had serious biases (Buolamwini and Gebru 2018). These machine-learning algorithms discriminated based on classes such as race and gender. Among their findings was the result that darker-skinned women are the most misclassified group. Their error rate rose to 34.7%, while the maximum error rate for light-skinned men was 0.8%.

Perhaps you can see why we mention this important research. If self-driving cars are trained in areas with a certain population, it is quite possible that they will also show a bias, such as recognizing people based on their skin color. If such a bias were the reason that self-driving cars significantly endangered people of a certain race or gender, say, there would obviously be very strong moral and legal reasons not to allow them on public roads. A recent study on the fairness of automated pedestrian detection does indeed find bias. Children in particular, but also women, are less likely to be detected (Li et al. 2023). Of course, this is a huge moral and legal problem, and our arguments are based on the assumption that self-driving cars do not introduce such new problems. Thus, our first premise for the above statement is that no other harms need to be introduced.

Another very brief example needs to be made explicit. Privacy. Self-driving cars rely on advanced technology that involves large amounts of data. Most likely, data will be connected in a network to enhance functionality, for example, to enable communication between cars. This may involve violations of privacy, which is also a protected human right. In the European Union, for example, the right to privacy, to control information about oneself, is essential to an autonomous life in dignity. This time, autonomy is actually meant in its traditional philosophical sense. Only if you control the data about yourself are you able to govern your life, so to speak. Therefore, new features of self-driving cars, such as commodities, must not be introduced at the expense of privacy.

1.7 The First Law of the Ethics of Self-Driving Vehicles: Reduce Risk

Unfortunately, self-driving vehicles will not completely eliminate fatal accidents. At least for a certain transition period, such vehicles will share roads and the larger public space with people whose behavior is error-prone and sometimes unpredictable. There will still be people getting on and off self-driving buses or crossing the road, say. For quite a long time there will also be mixed traffic, where self-driving cars have to share the road with human drivers, cyclists, and pedestrians. Human error will, thus, still be an issue for motorized traffic. Moreover, there will be technical errors. Sensors might fail and the software might have bugs, for example. Both the German Ethics Commission and the Expert Group on Specific Ethical Issues Raised by Driverless Mobility, advising the European Commission, recommend that self-driving vehicles should be programmed in a way that reduces risk

(European Commission Expert Group 2020; Ethics Commission 2017). That is, whenever possible, they will avoid dangerous situations by slowing down, swerving, or taking a different route. After all, given that self-driving vehicles will not only be automated but also 'connected,' they basically have all of the information about all of the other cars on the road, both past and present. That is, they know where the volume of traffic is particularly high and where (and why) accidents have happened in the past, etc. Given all this information, they can take you from A to B in the safest way possible, which may facilitate a shift toward pedestrian-oriented urban neighborhoods (Millard-Ball 2018).

Reducing risk is the predominant moral principle behind regulating self-driving vehicles. Take, for instance, a recent and comprehensive law regulating self-driving vehicles in Arizona, enacted in March 2021 and establishing standards for driverless vehicles.[13] This act is one of the first worldwide that does not distinguish between test mode and the normal operation of self-driving cars, as many preceding laws did. Commercial services like passenger transportation, freight transportation, and delivery operations, etc. can, thus, be offered. The car must comply with federal laws and standards as well as with all traffic and vehicle safety laws and, most interestingly for us, in case the automated driving system fails, it must 'achieve a minimal risk condition.'

This requirement is very understandable and complies, in a sense, with *Mill*'s principle briefly discussed above. And yet there is still a major problem. Given the technical dangers and the problems of mixed traffic, the only way to reduce the risk of accidents to zero would be to stop car traffic altogether. This will not happen. We have not banned and will not ban road traffic because there are fatal accidents. Rather, we accept some risks in order to enjoy the benefits of going by car. It is highly unlikely that this will change with the introduction of self-driving vehicles. There will be huge numbers of cars on the road, perhaps even more than today. It is thus to be expected that self-driving cars will also be involved in accidents, some of which will be fatal.

1.8 Moral Dilemmas with Self-Driving Vehicles

The innovation of self-driving vehicles presents us with the new, amazing, and, at the same time, arduous opportunity to decide in advance what should happen in which accident scenario. While accidents with conventional cars are often due to people's bad reactions—mostly fast and usually emotion-laden—we now have the chance to program the outcome of such accidents in a thoughtful, well-informed, and deliberate way.

This new opportunity forces us to think through the different kinds of accident scenarios self-driving vehicles might face. Some of these are pretty clear when viewed from a moral point of view. If all available options cause harm, the car should be directed to where it causes the least harm. For example, if an empty self-driving car can either hit a pedestrian (killing them) or a wall (damaging the car), it should clearly drive into the wall. And if it can either run over a pedestrian (killing them) or crash into another empty car, it should clearly crash into the car (damaging both cars). The German Ethics Commission, thus, stated that self-driving vehicles 'must be programmed to accept damage to animals or property in a conflict if this means that personal injury can be prevented' (Ethics Commission 2017, 11).

Some of the accident scenarios self-driving vehicles might face are more difficult to judge though. Imagine an empty self-driving car that can either hit a pedestrian, which would kill them, or a car with one person on board, which would kill that person. There are no easy answers to those moral dilemmas in which all available options lead to comparable harm (Sinnott-Armstrong 2005; McConnell 2024).

Therefore, the opportunity to decide in advance what should happen in which accident scenario also burdens us with having to decide who to harm and who to save. The arduous task involves deciding about moral dilemmas in advance. After all, it is not an option not to decide. Self-driving vehicles will have to be programmed in one way or the other. They will inevitably harm some people and save others. In the above-mentioned accident scenarios, there are but two options: The car can only proceed on its current course or swerve. The first option would result in one person being killed. The second option would result in damage to the car itself, damage to another car, or a passenger in the other car being killed. We can only decide whether it should stay on course or swerve.

You might think that there is no need to decide between the two options, though. After all, the car could be programmed to 'do nothing' in such dilemmas. 'Doing nothing' could be understood as the equivalent of drivers taking their hands off the steering wheel. Now what happens if they do that? The car probably proceeds according to the current position of the wheel. This might be straight ahead, but it might also not be. No matter how 'doing nothing' is understood, what attempts to shy away from making a decision actually do is substitute a direct answer to the moral question with an indirect one. Indirect answers such as 'hands off the wheel,' 'always keep in lane,' or 'decide randomly' rely on abstract rules suggesting that whatever happens is thanks to fate. Whatever happens is bad luck for the victim and good luck for those who were saved. This might sound tempting, but it is only an apparent solution to the moral problem. No matter how abstract and indirect the rule is, when included in programs for self-driving vehicles, it determines who they harm and who they save. Namely a human life here and a human life there. Either way, we bear the responsibility for our decision. If we decide not to program the car to swerve, this is a harmful and potentially blameworthy decision too, due to the mere fact that self-driving vehicles are vested with the possibility of being programmed for situations like that.

It is sometimes argued that discussing the dilemmas involving self-driving cars is not helpful and that we should focus on avoiding such dilemmas instead (European Commission Expert Group 2020, 17). We agree with the latter. Designing self-driving vehicles so that they cause fewer accidents and are involved in fewer dilemmatic accident scenarios is an incredibly important task. Hence, the first law for self-driving vehicles is to reduce harm.[14] This does not imply that they will never face such dilemma situations, however, or that it does not matter how they are programmed to deal with them, however seldom they might occur. Just recall the early accidents with the first automobiles. Fatal accidents can happen even at 4–8 mph. Hence, if we want to have perfectly safe self-driving cars, we must not allow anybody to come close to them—even up to the point that people are not allowed to ride in them. This is not a realistic option. Even today, we accept some risks in regular traffic. Staying at home to stay safe is not an option. Likewise, it is not an option to introduce a maximum speed limit for (self-driving) cars of 4–8 mph in order to make them perfectly safe. We should, thus, accept that self-driving cars will bring immense benefits while also bearing some risks. The question is, then, how self-driving vehicles should be programmed when all available options cause comparable harm.

1.9 Old Problems, New Challenges

The legal question of how to deal with dilemmas in road traffic is, in a sense, rather old. Thinking through imaginary cases is the bread and butter of legal scholars. As early as 1915, the legal scholar Josef Kohler presented the following hypothetical scenario with the title Car Driver case (*Autolenker-Fall*):

> Imagine that a car can no longer be brought to a standstill over a short distance, but that it is still possible to steer it in such a way that it drives to the right or to the left instead of moving straight ahead. If there are people straight ahead, on the right and on the left who can no longer avoid it, the driver is not in a position to avoid killing people, but he can divert death to one side or the other by moving the steering wheel. Can he be punished for causing the death of *A*, whereas if the car had continued in a straight line without moving the steering wheel, *B* or *C* would have been killed?
>
> *(Kohler 1915, 431–432, translation by the authors)*

What would you say? Is the driver guilty of directing the car toward *A* (in order to save *B* and *C*), thereby killing person *A*? He would have killed someone anyway, no matter what they did. Hence, does it make a difference who has to die? Philosophers and legal scholars have a long history of thinking about such cases. Take the 'Carneades case,' which goes back to Carneades of Cyrene (second-century BC): Two shipwrecked sailors see a plank that can only carry one of them. One of them reaches the plank first; when the other sailor also reaches the plank, both are quite exhausted. He pushes the first sailor off the plank, thus causing him to drown. Is the surviving sailor to blame or did he merely act in self-defense? If his behavior were forbidden, the law would effectively command sailors to sacrifice themselves. As this is too much to ask, such behavior, in German legal terms, is considered 'excusable'—i.e., it remains illegal but goes unpunished (Dreier 2007). In other words, the law recognizes the dilemma, a situation in which it is almost impossible to behave in a morally acceptable way.[15]

While, for a long time, thinking about similar cases was a theoretical exercise for philosophers, it was especially the dawn of the automobile age that made thinking about such cases of practical relevance for everyone. Kohler's intention as a criminal lawyer was to discuss whether the action chosen by the car driver was *punishable* under criminal law or whether emergency law ('Notrecht' in German) prevented the punishment of an individual who had no choice but to kill someone. Section 32 of the German Criminal Code regulates self-defense. You are not punished if you kill another person in order to save yourself or someone else.. Hence, if the driver in Kohler's case had been in danger (due to an illegal attack), it would not have been unlawful to kill the pedestrian. Section 33 (excessive self-defense) extends the justification and allows the killing of a person to 'whoever exceeds the limits of self-defense due to confusion, fear, or fright.' Do these provisions provide relief for the driver in our case? Arguably not, because he was not under 'unlawful attack.' According to these provisions, the driver would still have to be punished when he purposely redirected the car (in order to save someone else who was in danger).

Both sections require an attack as self-defense, which, according to Section 32 para 2, 'means any defensive action which is necessary to avert a present unlawful attack on oneself or another.' Therefore, these sections are actually rarely relevant for our cases as the pedestrians (B or C in Kohler's case) did not constitute an unlawful threat to the driver. They were simply in the wrong place (on the road) at the wrong time.

Emergency law has two other sections, however: Section 34 (necessity as justification) and Section 35 (necessity as defense). In emergency situations (in the case that two lives are in danger but neither of the endangered persons is engaged in an illegal attack endangering the life of the other person), the sacrifice of human life is not justifiable. Yet an act resulting in saving person A and killing person B might—according to a progressive view—be justifiable out of necessity or, at least, be excusable (Hilgendorf 2018). Note, however, that this verdict changes if A, B, and C in Kohler's case are known to the driver. If he identifies his nasty mother-in-law and chooses to steer the car to run her over on purpose, his action is not excusable anymore, even though he did so to prevent the killing of other people.

Although the legal question of how to deal with dilemmas in road traffic is rather old, it changes when asked in the context of self-driving vehicles. As long as people have been driving cars, dilemmatic accident scenarios have had to be accepted as fate, so to speak, because affective, subconscious human decisions determine the way accidents happen (except for the rare case of identifying your nasty mother-in-law on the road) (Goodall 2014; Feldle 2018). This is ultimately why these drivers are not punished under criminal law. But unconscious reactions in a fraction of a second cannot be compared with conscious, reflected decisions. Once we have the chance to think about who to kill and who to save, we are responsible for that decision. If Kohler were alive today, he would probably not be asking whether the driver could be punished for causing the death of person *A*, but whether self-driving cars should be programmed to kill *A*, *B*, or *C*. Although the outcome remains the same, the mere fact that it is not a person but an autopilot driving the car makes it a novel and tremendously difficult problem. Robots, after all, are still not subject to the penal code.

The kind of dilemma situation we have described here does not only exist in our vivid imagination. To underpin this point, we need to talk about football for a moment. In the run-up to the European Championship in Germany 2024, a documentary about the German team's great triumph at the 2014 World Cup in Brazil was shown on television (Roschitz and Heidrich 2024). One of the players at the time, Benedikt Höwedes, recounts his memories of the most important matches, how his hair fell out from all the excitement, etc. He also talks about the training camp that the team held in the Tyrolean Alps before the 2014 tournament. During this training camp, a promotional film was shot for a sponsor of the German team. As the sponsor was Mercedes, it was about cars—fast cars.

At just under 60 mph, Formula 1 driver Nico Rosberg and DTM driver Pascal Wehrlein were racing up the winding, closed-off mountain road in two 600 hp Mercedes 'racing' cars. Football star Julian Draxler was sitting with Rosberg in the car in front and Höwedes with Wehrlein in the second car. Only 6 m separated the two cars. There was a slope on one side of the mountain road and a barrier on the other side to prevent spectators from getting onto the road. Suddenly, Rosberg saw two people standing in front of the barrier giving hand signals. He understood the hand signals to mean that they wanted to warn him. Perhaps there was a dangerous obstacle on the track. In any case, he immediately slammed on the brakes. The driver of the second car, Wehrlein, did not know what was going on when the first car suddenly braked. He had to react immediately. As the distance between the cars was so small, he was unable to stop in time. So there were only three options. He could steer the car down the slope, he could hit the car in front, or he could drive into the barrier. Going down the slope would probably have killed himself and his passenger. Hitting the first car would also probably have killed or at least seriously injured the two occupants (and harmed those in the first car). Apparently, Wehrlein did not know where the third option would lead. However, he did in fact 'choose' this option. The car hit the two people in front of the barrier and seriously injured them. The fact that no one was killed can be considered a lucky escape.

Despite this traumatic experience, Höwedes and Draxler went to Brazil and played an outstanding tournament. The German Football Association did not report on the accident in detail at the time (neither did Mercedes). It is only now, ten years later, that the parties involved have decided to share more details about this tragic event. One of the victims is still in need of care today; Mercedes has apparently paid compensation. By the way, the waving was not meant as a warning. The person just wanted to cheer on the celebrities speeding past.

This example shows what can go wrong with cars and that tragic dilemma situations really do occur. Wehrlein did not know what would happen if he steered into the barrier. But a self-driving car might have known. Where should it have gone?

The rise in self-driving vehicles is a gift and a burden at the same time. Such vehicles give us the chance to save many lives. However, they also force us to decide who has to die in order to save others in dilemma situations. This is a question that requires broad public debate. As any solution to the question should be acceptable to the majority of people, it is important to involve them alongside moral experts and policymakers as well.

1.10 Who Should Decide?

Christoph von Hugo, the manager of driver assistance systems and active safety at Mercedes-Benz, said in an interview at the 2016 Paris Motor Show that the company will program their level 4 and 5 self-driving vehicles such that they prioritize the safety of their owners rather than pedestrians, or, one might add, people who simply cannot afford a Mercedes.[16] (Currently, the company offers its level 3 autopilot system only on cars that sell for approx. $150,000 and more.) This thinking is dangerous as it puts a price tag on weighty decisions about life-and-death scenarios in road traffic. Especially pedestrians might get the short end of the stick as car manufacturers are likely to be more motivated to program cars that favor their drivers as they are the ones buying the car (Contissa, Lagioia, and Sartor 2017). The question of how to make decisions about moral dilemmas involving self-driving vehicles must, hence, be answered by the law. The state even has a legal duty to ensure that fatal risks are diminished (remember Article 2 of the ECHR). It cannot be left up to car manufacturers to decide whether and how they program their self-driving vehicles in moral dilemmas since, except for special circumstances, companies are not allowed to make life-or-death decisions.[17]

Like Mercedes-Benz, most car manufacturers are interested in safety questions and also address extreme situations. Yet, apart from Mercedes-Benz, most of them do not include moral dilemmas in their considerations (Martinho et al. 2021).[18] Who can blame them? The nature of a dilemma is that either way, the outcome spells trouble. Such negative consequences are hard to accept and probably do not contribute to selling cars. Hence, if regulation does not force producers to deal with this issue, they will likely avoid facing it. So far in the United States, individual states have taken on the primary role in regulating self-driving vehicles. None of the states regulates moral dilemmas, however. Yet, they should do so because otherwise important decisions about life and death will be made by car manufacturers and not by the legislator.

So why not let individuals decide how to program their own self-driving cars for dilemma situations? The answer is fairly simple. While people generally prefer self-driving vehicles that favor the 'public good,' in the sense that they are programmed to save as many people as possible, they would rather not buy a self-driving vehicle programmed to sacrifice its passengers for the greater good (Bonnefon, Shariff, and Rahwan 2016). That might make them refrain from purchasing and using self-driving cars. As a consequence, self-driving cars might turn into shelf warmers, and, thus, their promise of saving more lives due to improvements in safety would be wasted. This 'social dilemma' can, however, be pre-empted by legal regulation if a fairly balanced distribution of negative burdens stemming from self-driving vehicles themselves is made obligatory for any such vehicle. The advantages of self-driving vehicles would arguably outweigh the downside of a disadvantageous program for a specific accident. If people were allowed to program their cars themselves, the coordination advantages of these cars would probably vanish too: Being able to coordinate the behavior of self-driving vehicles is a major factor for their superior safety (Pölzler 2021).

Furthermore, the philosophers Jan Gogoll and Julian F. Müller argue that mandatory ethics settings are in the best interests of society and, thus, of all individuals, because allowing individual ethics settings would lead to a prisoner's dilemma. In their words, 'I want to be moral, but I do not want to be a sucker' (Gogoll and Müller 2017, 691). They argue that if personal ethics settings were allowed, this would lead to an unfavorable outcome as nobody would want to be the nice guy who uses altruistic settings to save others at their own cost if the general expectation was that nobody else would be that nice and would use egoistic settings to their own advantage.

Others (Millar 2014), however, hold that individual ethics settings should be possible, while Nyholm (2018) argues for a middle ground in the sense that 'certain general boundaries' should be set within which 'people should be permitted, say, to be more altruistic than others are expected to be, if that truly is what they sincerely wish.' This compromise might be acceptable, but only to the extent that moral dilemmas are covered by the law. After all self-driving cars will have a hard time being accepted by the public if nobody knows whether owners have programmed their cars to sacrifice you in order to avoid damage to their expensive Mercedes. Nevertheless, beyond such decisions, an individual might, indeed, be allowed to sacrifice themselves, even to a degree not required by the law. However, individuals would not be free to disadvantage others for their own safety.

1.11 Conclusion

We hope that this first chapter helped you to understand that the ethical and legal challenges posed by self-driving vehicles are real, novel, and unavoidable, and that they require an answer. Self-driving cars will represent a tremendous leap forward, if technological developments unfold over the next few years as promised by the automotive industry. Remarkable progress comes with a huge burden, however. It is our task, as a society, to decide how to program self-driving vehicles for moral dilemmas. It is actually the first time in human history that we have had to make such a decision for such a mundane thing as public traffic, that we have to decide who we prioritize to save and who we put at risk of dying. This is no easy decision, and, thus, a great challenge for us as a society. Yet, not making a decision is not an option either because technology gives us the opportunity to program self-driving vehicles in advance for such situations. Hence, deciding not to decide is a decision too.

If you agree with us on these points, the problem we face is pretty clear. It is now time to solve it. Chapters 2 and 3 are about guiding you in your thinking about possible solutions as we are convinced that it is important that we the people take an informed decision on how to deal with moral dilemmas caused by self-driving vehicles. In what follows, we will guide you through some new attempts in philosophy to deal with dilemmas with self-driving cars. In the process, we will revisit some more traditional dilemmas and philosophical ideas.

Notes

1 See, however, Fallon and O'Neill (2005) pointing out the first fatal accident with a steam carriage in a small town in the Irish Midlands in 1869 where 'Mary Ward, a celebrated microscopist, artist, astronomer and naturalist, fell from a steam carriage and died after crush injuries from its heavy iron wheels.'

2 This was the sad end of a dramatic life. If you happen to have an interest in historical real crime stories, you might have heard about Mr *Bliss* before. He became a widower in 1895, when his stepdaughter killed his wife Evelina, allegedly by putting arsenic in a clam chowder. The trial was one of the first reported heavily in the media (Patalong 2012; Eschner 2017). While the driver of the cab that hit Mr Bliss was arrested and tried for murder, too, he was later released as he had had no intention whatsoever of killing Henry Bliss. This tragic accident is also remembered by a plaque that aims at promoting 'safety on our streets and highways,' which can still be seen today. (If you want to get an impression of the chaotic road traffic of those times, there is an interesting movie released on 21 April 1906 from the perspective of a tram driver in San Francisco called 'A Trip Down Market Street' by Harry J. Miles.)

3 This is not to say that human error is the only or always the main factor. What it means is that the crash would not have happened if the human driver had not made a mistake, see Zipper (2021).

4 For the pointed moniker 'Auto-Auto,' see, for example, *Spiegel* issue 9/2016.

5 The environment could profit as well if automation is regulated in order to reduce traffic and the distance traveled. The posited advantage for the environment is not a given, however, to the extent that even if self-driving cars greatly increase road safety and, at the same time, achieve an inclusion effect to the extent that many more individuals can now use these vehicles (Fagnant and Kockelman 2015), a great increase in the volume of traffic can be expected (see also Brown (2020)). Potential developments that make self-driving cars more ecological are not necessarily linked to the control technology becoming independent. Increased mobility, in the sense of increased traffic volume, should, thus, also be legally regulated to the extent that this new technology should also be sustainable.

6 Note that these Waymo cars are still not level 5 cars, as they are only allowed to operate in limited geographic areas and under certain conditions.

7 For an illuminating take on similar dilemmas, see Tessman (2017).

8 We mostly use the term 'self-driving cars' and rather rarely use 'autonomous vehicles,' but when we do, we use them interchangeably in this book, although 'autonomous' might be a potentially misleading notion in this context because in philosophy and law the term 'autonomy' is typically taken very literally. As *autós* means self and *nómos* means law, autonomy is taken to refer to something like self-government—if a person is autonomous, they are their own lawmaker. That is certainly not the case for self-driving cars as they are designed to drive to a specific place, instead of deciding themselves where it would be nice to go. Their driving automation enables them to execute what somebody else has decided to do but they are by no means actually autonomous. Nevertheless, as both notions are frequently used in academic and public discourse alike, we consider it sufficient to clearly state that autonomous vehicles as we address them here are not self-governing. Having said this once, we are convinced that you will not be misled by the term

autonomous vehicle. Nevertheless, it is worth pointing out that the SAE's current classification of level 5 no longer includes the designation of such vehicles as 'autonomous.' Referring to cars at level 4, the German Act on Autonomous Driving 2021 still speaks of cars with 'an autonomous driving function,' which, according to Section 1d, is given if the car 'can independently perform the task of driving without a person driving the vehicle within a specified operating range.'

9 In fact, it is perhaps even more relevant for cyclists or pedestrians than for car owners because self-driving cars can be programmed to reduce current inequalities in vulnerability among road users. As cyclists are currently much more likely to be seriously harmed than car drivers when involved in traffic accidents, self-driving cars provide us with the opportunity to protect them from cars. They can be programmed to automatically slow down when a bicycle has been detected or to give them enough space, say (European Commission Expert Group 2020, 31).

10 See the ECtHR case *Öneryildiz v Turkey*, App no 69546/12 (ECtHR, 30 November 2004) para 71 where the court stated that Article 2 of the European Convention on Human Rights, the right to life, 'lays down a positive obligation on States to take appropriate steps to safeguard the lives of those within their jurisdiction' and this 'must be construed as applying in the context of any activity, whether public or not, in which the right to life may be at stake, and *a fortiori* in the case of industrial activities, which by their very nature are dangerous, such as the operation of waste-collection sites.' [italics original] Cf *Budayeva ea v Russia*, App nos 15339/02 ea (ECtHR, 20 March 2008) in relation to natural disasters. For an overview, see, e.g., Schabas (2015).

11 See Grabenwarter and Pabel (2021) with further references on *Nicolae Virgiliu Tănase v Romania* (GC), App no 41720/13 (ECtHR, 25 June 2019), para 135.

12 Cf., e.g., *Pretty v United Kingdom*, App no 2346/02 (ECtHR, 29 April 2002), para 39 'The consistent emphasis in all the cases before the Court has been the obligation of the State to protect life.' For the literature, see Eisenberger (2017).

13 HB 2813.

14 See Section 1.7.

15 Note that in German criminal law this recognition comes with the somewhat strange solution of illegal behavior not being punished. In English and Welsh criminal law, we find something similar in the defence 'balance of evil,' most recently established in the Court of Appeal case of the Conjoined Twins (Re A (Conjoined Twins: Surgical Separation) [2000] 4 All ER 961) and their surgical separation, which led to the death of one twin in order to save the life of the other, as being permissible out of necessity (we will discuss this case in Section 3.2). See, e.g., Ashworth and Horder (2013). Nevertheless, this is an exceptional case in English and Welsh criminal law with highly unusual facts, see Herring (2012). For the U.S. context, with a discussion of Trolley cases, see Dressler and Garvey (2019). While the nitty-gritty legal details are complicated, the general recognition of the moral dilemma is shared.

16 See, e.g., Taylor (2016) providing quotations of what Mr von Hugo said. For example, 'If you know you can save at least one person, at least save that one. Save the one in the car,' or 'You could sacrifice the car. You could, but then the people you've saved initially, you don't know what happens to them after that in situations that are often very complex, so you save the ones you know you can save.' But see, too, Bonnefon (2021) who provides the context for this supposedly outrageous statement.

17 Cf. *Pretty v United Kingdom*, App no 2346/02 (ECtHR, 29 April 2002), para 37ff; see also Eisenberger (2017). For a discussion on possible individual ethics settings, see Kirchmair (2023) with further references.

18 The only company which comes close to describing what its self-driving cars would do in the case of a moral dilemma is Nuro, a company specializing in automating delivery technology, which stated that in extreme cases they would prioritize humans over their content (their vehicles do not have passengers).

References

Ashworth, Andrew, and Jeremy Horder. 2013. *Principles of Criminal Law*. 7th ed. Oxford: Oxford University Press.

Bond, Shannon. 2018. "Uber Resumes Autonomous Vehicle Testing." *Financial Times*, December 20, 2018. www.ft.com/content/771300d6-03ce-11e9-99df-6183d3002ee1.

Bonnefon, Jean-François. 2021. *The Car That Knew Too Much: Can a Machine Be Moral?* Cambridge, MA: MIT Press.
Bonnefon, Jean-François, Azim Shariff, and Iyad Rahwan. 2016. "The Social Dilemma of Autonomous Vehicles." *Science* 352 (6293): 1573–76. https://doi.org/10.1126/science.aaf2654.
Brink, David. 2007. Mill's Moral and Political Philosophy. https://plato.stanford.edu/entries/mill-moral-political/#HarPri.
Brown, Austin. 2020. "Demanding a Better Transportation Future through Automation." In *Are We There Yet?: The Myths and Realities of Autonomous Vehicles*, edited by Michael A. Pagano, P. S. Sriraj, Austin Brown, Stan Caldwell, Chris T. Hendrickson, Kazuya Kawamura, and Taylor Long, 25–46. Urbana: University of Illinois Press.
Buolamwini, Joy, and Timnit Gebru. 2018. "Gender Shades: Intersectional Accuracy Disparities in Commercial Gender Classification." In *Proceedings of the 1st Conference on Fairness, Accountability and Transparency*, 77–91. PMLR. https://proceedings.mlr.press/v81/buolamwini18a.html.
Chan, Mick. 2021. "Mercedes-Benz Drive Pilot Cleared for Use in Germany." *Paultan.Org*. https://paultan.org/2021/12/10/mercedes-benz-drive-pilot-cleared-for-use-in-germany/.
Contissa, Giuseppe, Francesca Lagioia, and Giovanni Sartor. 2017. "The Ethical Knob: Ethically-Customisable Automated Vehicles and the Law." *Artificial Intelligence and Law* 25 (3): 365–78. https://doi.org/10.1007/s10506-017-9211-z.
Deutsches Patent- und Markenamt. 2022. "Bertha Benz." January 1, 2022. www.dpma.de/english/our_office/publications/ingeniouswomen/berthabenz/index.html.
Dingus, Thomas A., Feng Guo, Suzie Lee, Jonathan F. Antin, Miguel Perez, Mindy Buchanan-King, and Jonathan Hankey. 2016. "Driver Crash Risk Factors and Prevalence Evaluation Using Naturalistic Driving Data." *Proceedings of the National Academy of Sciences* 113 (10): 2636–41. https://doi.org/10.1073/pnas.1513271113.
Dreier, Horst. 2007. "Grenzen des Tötungsverbotes – Teil 1." *JuristenZeitung* 62 (6): 261–70. https://doi.org/10.1628/002268807780282749.
Dressler, Joshua, and Stephen P. Garvey. 2019. *Criminal Law: Cases and Materials*. 8th ed. American Casebook Series. St. Paul, MN: West Academic Publishing.
Eisenberger, Iris. 2017. "Das Trolley-Problem im Spannungsfeld autonomer Fahrzeuge: Lösungsstrategien grundrechtlich betrachtet." In *Autonomes Fahren und Recht*, edited by Iris Eisenberger, Konrad Lachmayer, and Georg Eisenberger, 91–107. Wien: MANZ Verlag.
Eschner, Kat. 2017. "Henry Bliss, America's First Pedestrian Fatality, Was Hit By an Electric Taxi." *Smithsonian Magazine*, September 13, 2017. www.smithsonianmag.com/smart-news/henry-bliss-americas-first-pedestrian-fatality-was-hit-electric-taxi-180964852/.
Ethics Commission. 2017. "Report on Automated and Connected Driving." June 1, 2017. www.bmvi.de/SharedDocs/EN/publications/report-ethics-commission.pdf?__blob=publicationFile.
European Commission Expert Group. 2020. "Ethics of Connected and Automated Vehicles: Recommendations on Road Safety, Privacy, Fairness, Explainability and Responsibility." Brussels. https://op.europa.eu/o/opportal-service/download-handler?identifier=89624e2c-f98c-11ea-b44f-01aa75ed71a1&format=pdf&language=en&productionSystem=cellar&part=.
Fagnant, Daniel J., and Kara Kockelman. 2015. "Preparing a Nation for Autonomous Vehicles: Opportunities, Barriers and Policy Recommendations." *Transportation Research Part A: Policy and Practice* 77 (January): 167–81. https://doi.org/10.1016/j.tra.2015.04.003.
Fallon, Isabelle, and Desmond O'Neill. 2005. "The World's First Automobile Fatality." *Accident Analysis & Prevention* 37 (4): 601–3. https://doi.org/10.1016/j.aap.2005.02.002.
Feldle, Jochen. 2018. *Notstandsalgorithmen: Dilemmata im automatisierten Straßenverkehr*. Baden-Baden: Nomos.
Gogoll, Jan, and Julian F. Müller. 2017. "Autonomous Cars: In Favor of a Mandatory Ethics Setting." *Science and Engineering Ethics* 23 (3): 681–700. https://doi.org/10.1007/s11948-016-9806-x.

Goodall, Noah J. 2014. "Machine Ethics and Automated Vehicles." In *Road Vehicle Automation*, edited by Gereon Meyer and Sven Beiker, 93–102. Lecture Notes in Mobility. Cham: Springer International Publishing. https://doi.org/10.1007/978-3-319-05990-7_9.

Google Walkout For Real Change. 2020. "Standing with Dr. Timnit Gebru—#ISupportTimnit #BelieveBlackWomen." *Medium* (blog), December 15, 2020. https://googlewalkout.medium.com/standing-with-dr-timnit-gebru-isupporttimnit-believeblackwomen-6dadc300d382.

Grabenwarter, Christoph, and Katharina Pabel. 2021. *Europäische Menschenrechtskonvention: Ein Studienbuch*. 7th ed. Munich; Vienna; Basel: C.H. Beck; Manz; HLV.

Herring, Jonathan. 2012. *Criminal Law: Text, Cases, and Materials*. 5th ed. Oxford: Oxford University Press.

Hilgendorf, Eric. 2018. "The Dilemma of Autonomous Driving: Reflections on the Moral and Legal Treatment of Automatic Collision Avoidance Systems." In *Digitization and the Law*, edited by Eric Hilgendorf and Jochen Feldle, 57–90. Baden-Baden: Nomos.

Honda. 2021. "Honda to Begin Sales of Legend with New Honda SENSING Elite | Honda Global Corporate Website." *Honda Global*. https://global.honda/en/newsroom/news/2021/4210304eng-legend.html.

Information Britain. 2010. "The 12th of February 1898 AD, Britains 1st Car Crash Fatality." www.information-britain.co.uk/famdates.php?id=877.

Jones, Rachyl. 2024. "Exclusive: Mercedes Becomes the First Automaker to Sell Autonomous Cars in the U.S. That Don't Come with a Requirement That Drivers Watch the Road." *Fortune*. https://fortune.com/2024/04/18/mercedes-self-driving-autonomous-cars-california-nevada-level-3-drive-pilot/.

Kirchmair, Lando. 2023. "How to Regulate Moral Dilemmas Involving Self-Driving Cars: The 2021 German Act on Autonomous Driving, the Trolley Problem, and the Search for a Role Model." *German Law Journal* 24 (7): 1184–1208. https://doi.org/10.1017/glj.2023.83.

Kohler, Josef. 1915. "Das Notrecht." *Archiv für Rechts- und Wirtschaftsphilosophie* 8 (4): 411–49.

Levin, Sam, and Julia Carrie. 2018. "Self-Driving Uber Kills Arizona Woman in First Fatal Crash Involving Pedestrian." *The Guardian*. www.theguardian.com/technology/2018/mar/19/uber-self-driving-car-kills-woman-arizona-tempe.

Li, Xinyue, Zhenpeng Chen, Jie Zhang, Federica Sarro, Y. Zhang, and Xuanzhe Liu. 2023. "Bias Behind the Wheel: Fairness Analysis of Autonomous Driving Systems." www.semanticscholar.org/paper/Bias-Behind-the-Wheel%3A-Fairness-Analysis-of-Driving-Li-Chen/2d50c18af29a3709b996a8bd905596630512e21f.

Litman, Todd. 2020. "Autonomous Vehicle Implementation Predictions: Implications for Transport Planning." Transportation Research Board. https://trid.trb.org/View/1678741.

London Remembers. 2024. "First Driver Killed in Road Accident." *London Remembers*. www.londonremembers.com/memorials/first-driver-killed-in-road-accident/.

Martinho, Andreia, Nils Herber, Maarten Kroesen, and Caspar Chorus. 2021. "Ethical Issues in Focus by the Autonomous Vehicles Industry." *Transport Reviews* 41 (5): 556–77. https://doi.org/10.1080/01441647.2020.1862355.

McConnell, Terrance. 2024. "Moral Dilemmas." In The Stanford Encyclopedia of Philosophy, edited by Edward N. Zalta and Uri Nodelman, Spring 2024. Metaphysics Research Lab, Stanford University. https://plato.stanford.edu/archives/spr2024/entries/moral-dilemmas/.

McFarlane, Andrew. 2010. "How the UK's First Fatal Car Accident Unfolded." *BBC News*, August 17, 2010, sec. Magazine. www.bbc.com/news/magazine-10987606.

Mill, John Stuart. 1965. "On Liberty." In *The Collected Works of John Stuart Mill*, edited by J. M. Robson, 213–310. Toronto: University of Toronto Press. https://oll.libertyfund.org/title/robson-the-collected-works-of-john-stuart-mill-volume-xviii-essays-on-politics-and-society-part-i.

Millar, Jason, ed. 2014. *Technology as Moral Proxy: Autonomy and Paternalism by Design*. Chicago: IEEE.

Millard-Ball, Adam. 2018. "Pedestrians, Autonomous Vehicles, and Cities." *Journal of Planning Education and Research* 38 (1): 6–12. https://doi.org/10.1177/0739456X16675674.

NHTSA. 2022. "Summary Report: Standing General Order on Crash Reporting for Automated Driving Systems." National Highway Traffic Safety Administration. www.nhtsa.gov/sites/nhtsa.gov/files/2022-06/ADS-SGO-Report-June-2022.pdf.

Nyholm, Sven. 2018. "The Ethics of Crashes with Self-Driving Cars: A Roadmap, I." *Philosophy Compass* 13 (7): e12507. https://doi.org/10.1111/phc3.12507.

Papadoulis, Alkis, Mohammed Quddus, and Marianna Imprialou. 2019. "Evaluating the Safety Impact of Connected and Autonomous Vehicles on Motorways." *Accident Analysis & Prevention* 124 (January): 12–22. https://doi.org/10.1016/j.aap.2018.12.019.

Patalong, Frank. 2012. "Das Ende des Henry Hale Bliss." *Spiegel Geschichte* 18. www.spiegel.de/geschichte/historische-verkehrsunfaelle-a-947763.html.

Pölzler, Thomas. 2021. "The Relativistic Car: Applying Metaethics to the Debate about Self-Driving Vehicles." *Ethical Theory and Moral Practice*, May. https://doi.org/10.1007/s10677-021-10190-8.

Porter, A. 1998. "First Fatal Car Crash in Britain Occurred in 1898." *BMJ (Clinical Research Ed.)* 317 (7152): 212.

Prawitz, Sven. 2023. "Mercedes-Benz testet Level-3-System in Peking und plant Auslieferung in den USA." *»Automobil Industrie«*. www.automobil-industrie.vogel.de/mercedes-benz-assistenzsystem-level-3-peking-test-a-360a820f01fab76f34d5fa58499a16b8/.

Roschitz, Martin, and Matthias Heidrich. 2024. "WM-Triumph 2014: 'Dann bist du mausetot' – Höwedes und die Katastrophe vor dem Titel." NDR.de. 2024. www.ndr.de/sport/fussball/WM-2014-Dann-bist-du-mausetot-Hoewedes-und-die-Katastrophe-vor-dem-Titel,wm1750.html.

SAE International. 2021. "Taxonomy and Definitions for Terms Related to Driving Automation Systems for On-Road Motor Vehicles." 2nd ed. www.sae.org/standards/content/j3016_202104/.

Schabas, William. 2015. *The European Convention on Human Rights: A Commentary*. Oxford: Oxford University Press. www.loc.gov/catdir/enhancements/fy1604/2015937610-b.html.

Shepardson, David. 2023. "Backup Driver in 2018 Uber Self-Driving Crash Pleads Guilty." Reuters, July 28, 2023, sec. Autos & Transportation. www.reuters.com/business/autos-transportation/backup-driver-2018-uber-self-driving-crash-pleads-guilty-2023-07-28/.

Sinnott-Armstrong, Walter. 2005. "Moral Dilemmas." In *Encyclopedia of Philosophy*. www.encyclopedia.com/humanities/encyclopedias-almanacs-transcripts-and-maps/moral-dilemmas.

Somerville, Heather, and David Shepardson. 2018. "Uber Car's 'safety' Driver Streamed TV Show before Fatal Crash: Police." *Reuters*. www.reuters.com/article/us-uber-selfdriving-crash-idUSKBN1JI0LB.

Sparrow, Robert, and Mark Howard. 2017. "When Human Beings Are Like Drunk Robots: Driverless Vehicles, Ethics, and the Future of Transport." *Transportation Research Part C: Emerging Technologies* 80 (January): 206–15. https://doi.org/10.1016/j.trc.2017.04.014.

Taylor, Michael. 2016. "Self-Driving Mercedes-Benzes Will Prioritize Occupant Safety over Pedestrians." Car and Driver. www.caranddriver.com/news/a15344706/self-driving-mercedes-will-prioritize-occupant-safety-over-pedestrians/.

Tessman, Lisa. 2017. When Doing the Right Thing Is Impossible. *Philosophy in Action*. New York, NY: Oxford University Press.

The Guardian. 2014. "From the Archive, 26 August 1896: The UK's First Fatal Car Accident," August 26, 2014, sec. World news. www.theguardian.com/world/2014/aug/26/uk-first-fatal-car-accident-archive-1896.

Vitzthum, Thomas. 2017. "Als Merkel in die Zukunft blicken soll, lacht das Auditorium." *Welt*, June 9, 2017. www.welt.de/politik/deutschland/article165359594/Als-Merkel-in-die-Zukunft-blicken-soll-lacht-das-Auditorium.html.

Waymo. 2023. "Waymo Significantly Outperforms Comparable Human Benchmarks over 7+ Million Miles of Rider-Only Driving." Waymo. https://waymo.com/blog/2023/12/waymo-significantly-outperforms-comparable-human-benchmarks-over-7-million.

Winton, Neil. 2022. "Computer Driven Autos Still Years Away Despite Massive Investment." *Forbes*, February 27, 2022. www.forbes.com/sites/neilwinton/2022/02/27/computer-driven-autos-still-years-away-despite-massive-investment/?sh=4ffec4a718cc.

World Health Organization. 2018. "Global Status Report on Road Safety 2018." www.who.int/publications/i/item/9789241565684.

Zipper, David. 2021. "The Deadly Myth That Human Error Causes Most Car Crashes." *The Atlantic*, November 26, 2021. www.theatlantic.com/ideas/archive/2021/11/deadly-myth-human-error-causes-most-car-crashes/620808/.

2

HOW TO REGULATE MORAL DILEMMAS INVOLVING SELF-DRIVING CARS

2.1 The Basics: Reduce Harm! Prioritize Human Lives!

It is to be expected that self-driving vehicles will be involved in accidents. A small proportion of these will be dilemmas in the sense that people being killed—be they occupants, cyclists, or pedestrians—is unavoidable. In 2021, Germany enacted the first law worldwide that regulates the programming of self-driving vehicles, including provisions for dilemma situations.

The German Act on Autonomous Driving says that self-driving cars must have an accident-avoidance system that

a is designed to avoid and reduce harm;
b in the event of unavoidable alternative harm to different legal interests, takes into account the importance of the legal interests, with the protection of human life having the highest priority; and
c in the case of unavoidable alternative harm to human life, does not provide for further weighting on the basis of personal characteristics.[1]

This law is largely based on a report by the German Ethics Commission mentioned above (Ethics Commission 2017). The Ethics Commission formulated 20 quite precise rules for parliament to consider when drafting the law. Rule No. 2, for example, states that cars should be programmed such that accidents will be avoided whenever this is possible. Rule No. 7 demands programming that strictly prioritizes human life over that of animals and over material damage, because nothing is as important and as valuable as a human life. So whenever a car can save people by running into a dog or crashing into a wall, say, it should do so, no matter how significant the material damage is. These two rules were taken over directly in the law, which is not surprising since they are so obvious that even back in

DOI: 10.4324/9781003485216-2

1915 the legal scholar Josef Kohler, who we met in Chapter 1, deemed them to be clear. He wrote that

> [t]he causal movement, which in case of emergency aims at one or another human being, cannot make the driver of the car responsible; while it would be different, of course, if the driver of the vehicle could have driven into a flock of sheep and thus saved human lives.
> *(Kohler 1915, 432, translation by the authors)*

2.2 Personal Characteristics Don't Matter! Or Do They?

For dilemma situations, in which all options inevitably lead to the loss of human life, the Ethics Commission recommended prohibiting any kind of qualification according to personal characteristics such as age, sex, and physical or mental constitution. That is, the cars must not be programmed to save children instead of the elderly, women instead of men, or the healthy instead of the sick (Rule No. 9). This recommendation, which led to lit. c in the German law above, arguably also includes a prohibition of preferential treatment for members of certain ethnicities or faiths.

These three points—minimization of harm, prioritization of human lives, and the prohibition of qualifying human lives according to personal characteristics—seem to be the bare minimum that should not be left up to individuals or companies to decide. No matter how much you pay for your Mercedes, this does not buy you the right to program your car such that it prefers to save a person of a certain gender, skin tone, or age. At least these elements of programming seem to be off the table for individual ethics settings. This is also the position represented by our fictional friend, the clear-headed philosopher, Professor *Principia*. She holds that the German Ethics Commission had good reasons for drafting their rules and that the German lawmakers rightly adopted them.

But not so fast, a critical voice interjects. This is our second fictional friend, the witty philosopher, Professor Skeptico. Skeptico, too, has read the work of Kohler. He remembers that Kohler posed the question as to whether it would be acceptable to rank human beings according to their importance:

> One could perhaps try a scale of values in such a way that persons and classes of people were divided into different categories according to their importance for the world culture. Such an attempt was made by earlier legal systems: think of the Indian caste system.

Professor Principia hastens to clarify that Kohler did, indeed, speculate about such an idea but was wise enough to quickly dismiss it when he concluded his initial thought by stating that:

> [i]t is no longer possible in our day: we cannot make an exception to the protection of the individual, nor can we treat a person of lower rank less than a person of higher rank, nor can we treat a person of a lower race less than a man of culture.[2]

One might think, Principia adds, that moral progress over the past 100 years would have ensured that such things are no longer possible today.

But Skeptico is not convinced. He is also up to date in the social sciences. Just over 100 years later, in 2016, researchers dared to ask a question very similar to Kohler's. But their question was not merely rhetorical; instead, it was part of a large-scale experiment (the so-called 'Moral Machine Experiment') that aimed to receive a concrete answer to moral dilemmas with self-driving vehicles. A total of 2.3 million people in 233 countries or regions took part, which is an incredibly large data set. This research team found that surprisingly, or rather shockingly, people around the globe—with some minor regional differences—seem to have a strong preference to save younger over older lives, a somewhat weaker but still marked tendency to spare people with a higher social status (e.g., a doctor in contrast to a criminal, or a business executive vs. a homeless person), and women before men (Awad et al. 2018).

While—in contrast to Kohler—the Moral Machine Experiment did not suggest categorizing people into caste-like groups comprising a certain religion or ethnicity for the purpose of ranking them, the researchers, nevertheless, used social status, age, and gender as potentially decisive characteristics which could lead to individuals being spared in accidents with self-driving cars, Skeptico explains.

These findings, which were published in the journal *Nature* (perhaps the most important academic journal there is), stirred a heated debate. So let us look at these findings in more detail. We will discuss social status, age, and gender in turn.

2.3 Is It Right to Prioritize Doctors over Criminals?

Principia also did her homework, reading up on the somewhat adventurous story of how the researchers behind the Moral Machine Experiment came up with the idea for the experiment. This story was written by one of the architects of the experiment, the French psychologist Jean-François Bonnefon, a few years after the article was published. Principia was relieved to read in *Bonnefon*'s book that the category of social status was meant to show the 'limits of an exercise such as Moral Machine for informing public policy.' For the researchers, '[c]itizens' preferences are not always acceptable from an ethical point of view, and the purpose of Moral Machine is therefore not to replace ethics committees' (Bonnefon 2021, 115). So, apparently, they did regard it as obvious that social status should not be a basis for differential treatment in the programming of self-driving vehicles. Their intention to include it as a category in their experiment in the first place was to highlight the limits of their approach.

But even if we grant them a certain naivety in their desire to act with the best of intentions, they unleashed the devil, Skeptico replies. If there is a statistically significant preference for saving doctors over criminals, maybe we have to reconsider the standard narrative of equality holding that we ought to treat all people equally no matter what their status in society is. One might well take the social status preference found in the Moral Machine Experiment and argue that it is more about individualistic than native group-based discrimination. While people are born into a certain social class like a caste, they are not born as a doctor or a business executive, or as a potential criminal or homeless person.

To understand this thought, let us take a look at the design of the Moral Machine Experiment. As you will see when you go to *moralmachine.mit.edu* (yes, the website is still

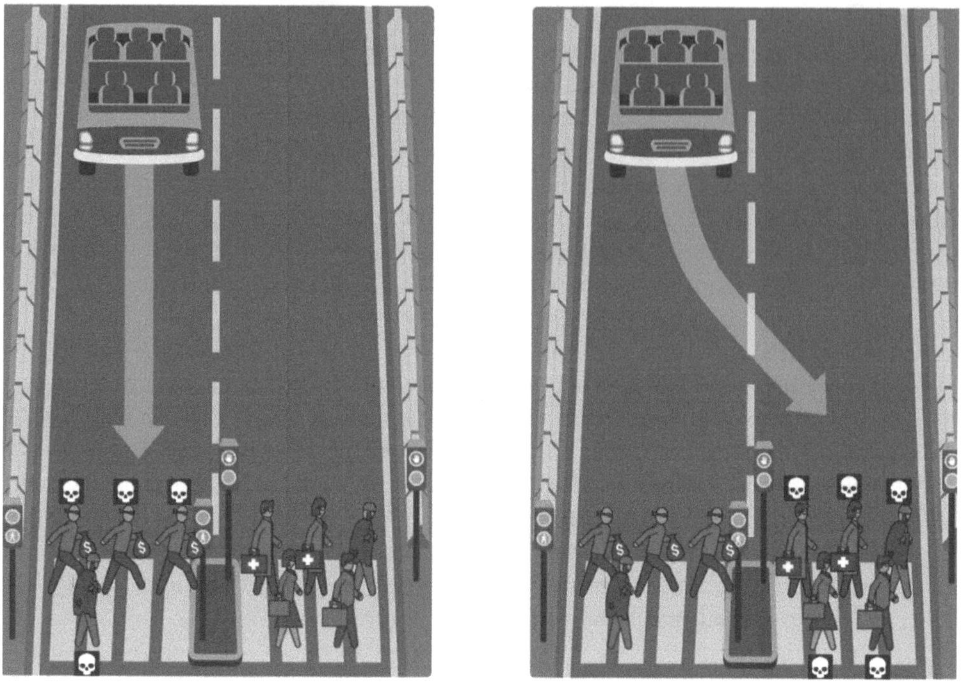

FIGURE 2.1 An image from the Moral Machine Experiment asking participants to decide between killing doctors, killing criminals, and killing homeless people.

active: go and have a look for yourself), the Moral Machine Experiment is an online tool that looks a bit like an old-fashioned computer game. When you participate in the experiment, you are confronted with a series of dilemma situations, one by one. Each situation is shown as a relatively simple static picture featuring an out-of-control car and little persons who are in danger of being hit by the car (see Figure 2.1). What changes from scenario to scenario is where the car can go and which person will be hurt or killed depending on where the car is going. The persons differ in their personal appearances. As you can see in Figure 2.1, an image from the website, some of the persons are carrying a red bag displaying a white cross (signaling that this is a doctor) while others have a bag of loot (signaling that this is a criminal) or a coat with holes (signaling that this is a homeless person), for example. One can also click on a picture description, which clarifies that one person is a doctor and the other is a criminal.

As a participant in the Moral Machine Experiment, your task is to make a decision in various dilemma situations. That is, you have to decide whether the car should stay on course and kill the doctor or if it should swerve to the left and kill the criminal in order to save the doctor, say. What is important to note here, Professor Skeptico explains, is that participants always decide what the car should do in specific dilemma situations. Never was there a box to tick that said something as abstract as 'spare someone with a higher social status.'

This is important because the participants might well have had something more noble than social status in mind, Skeptico says. For instance, they might have reckoned that society benefits from doctors because they cure people's illnesses and prevent deaths whereas criminals harm society. Similarly, they might have regarded business executives as being good for society by contributing to its wealth whereas homeless people do not usually do so. In other words, Skeptico says, maybe the participants in the Moral Machine Experiment did not rank the figures according to some abstract notion of social status. Instead, they might have evaluated the benefit a particular figure brings to society.

Principia calmly responds that this is all well and good but pure fantasy. In the real world, she says, doctors rarely cross the street carrying a red doctor's bag (let alone wearing a red coat and trousers). And criminals rarely have a bag of loot with them (with a Dollar sign on it) and a somewhat mysterious headband.[3] Self-driving vehicles will hardly be able to detect who is a doctor and who is a criminal.

Wait a minute, Skeptico says, this is not fiction. Use your imagination. It is not far-fetched to assume that enough information will be available to highly connected self-driving cars in the not-too-distant future. As pretty much every person has some kind of device with them that is permanently connected to the internet (smartphone, smartwatch, etc.), it is easy to imagine that such devices would send a signal to all self-driving cars that this person is a doctor, say. Similarly, these devices might be used to indicate who is a convicted criminal—or whatever you might find relevant. To a certain extent, this is already current practice: convicted sex offenders are listed in public registries or databases in the United States and the United Kingdom, for example. Such devices might also be used to protect previously vulnerable groups. For instance, they could signal that a person has certain special needs, prompting self-driving cars to automatically slow down and give them more space than other people.

But Principia insists that re-interpreting the finding of the Moral Machine Experiment regarding social status in terms of its benefit to society is perhaps a smart idea but not a very ethical one because it still treats people differently depending on how beneficial they are. That is a clear-cut violation of the promise that all people are born equal, she snorts angrily. People's value, she says, does not depend on how beneficial they are. They are not like shoes that are thrown away once they are worn out. People have an inherent value, human dignity, if you will. That seems to be the major reason why the research team surrounding the Moral Machine Experiment and many more philosophers, lawyers, and like-minded people find that social status or an individual's value to society should not be a ground for differential treatment in the programming of self-driving vehicles.

However, Professor Skeptico is not convinced and not prepared to give up that fast. The fact that one person is somehow more beneficial to society than another person has been used as a ground for preferential treatment in many contexts. And we are not talking about salary gaps between the world's most famous football player and a humble philosophy teacher. He offers yet another, more recent example. Think of triage situations in a hospital's emergency unit during the COVID-19 pandemic. Imagine the following situation. Two patients arrive at the emergency unit at exactly the same time; both of them need a respirator to survive but only one respirator is available. Who should get it (and survive) and who should not get it (and die)?

As long as we have no information about the two patients, it seems impossible to find a moral solution to this dilemma, Skeptico holds. Both deserve the treatment, but it is not available to both. But what if one of the patients is a doctor, a much-needed lung expert, he suggests, and the other one is a philosopher who teaches ethics at university, or, even worse, a lawyer? This difference seems to be relevant, perhaps even from a moral point of view. If the doctor is cured, they can help fight the pandemic, potentially saving many lives; if the philosopher is cured, they can, at best, write a book about triage situations, while the lawyer might sue the hospital despite being spared. This does not, of course, imply, he hastens to add, that philosophers or lawyers do not have inherent value. It just means that there might be features that are even more relevant from a moral point of view.

2.4 How We Learned to Sort People: Triage

Consider the origins of what we know as 'triage' today, Professor Skeptico says. They date back to 1792, when Baron Dominique Jean Larrey, Surgeon-in-Chief to Napoleon's Imperial Guard, was tasked with managing casualties in the French Army (Robertson-Steel 2006). The French verb *trier* means 'to separate out,' in the sense of sorting items according to quality. The goal was to quickly return wounded soldiers to active service. Hence, a soldier's expected performance on the battlefield was the criterion used to decide who would be 'separated out' for survival (and who was not), which sounds a lot like an old version of the benefit-to-society reasoning.

This is not the whole story, Professor Principia argues. While it does seem to be the case that triage goes back to Baron Larrey, he was also inspired by the French Enlightenment, as apparent in one of his rules for sorting soldiers described in his memoirs on the Russian campaign in 1812:

> Those who are dangerously wounded should receive the first attention, without regard to rank or distinction. They who are injured in a less degree may wait until their brethren in arms, who are badly mutilated, have been operated on and dressed, otherwise the latter would not survive many hours; rarely, until the succeeding day.[4]

True, Skeptico concedes. But this is not the whole story either. In 1846, the British naval surgeon John Wilson suggested yet another rule to sort soldiers. He argued that the focus should be on those who need immediate treatment that is likely to be successful. In other words, what is most important is not the severity of the injury alone but the severity combined with the likelihood of recovery. This means that those with wounds that are likely to be fatal should not be treated (Iserson and Moskop 2007).

In World War I yet another way emerged to think about who to select for care. As W. W. Keen wrote in 1917:

> [A] hospital with 300 or 400 beds may suddenly be overwhelmed by 1000 or more cases (think of a train accident, for instance). It is often, therefore, physically impossible to give speedy and thorough treatment to all. A single case, even if it urgently requires attention—if this will absorb a long time—may have to wait, for in that same time a dozen others, almost equally exigent, but requiring less time, might be cared for. The greatest good of the greatest number must be the rule.
>
> *(Iserson and Moskop 2007)*

The COVID-19-pandemic provides a more recent version of the same idea. Very early in 2020, Italy was the first Western state to be hit hard by COVID-19 before it became a pandemic. In this situation, the Italian Society of Anaesthesia, Analgesia, Resuscitation, and Intensive Care issued new guidelines to intensive care physicians for situations in which they were no longer able to provide ventilators to all those who needed them. The recommendation was that scarce resources should be primarily reserved for those who are more likely to survive; the second priority is to treat those who can gain more life years; the general aim is to maximize the benefit for the greatest number of people (Lübbe 2022).

This reminds Professor Skeptico of another aspect of this most recent pandemic: Back in 2021, when the first vaccines against COVID-19 became available but were still scarce, most countries (if not all) decided to distribute them according to the vulnerability of people *and* according to their importance to society. You would get a shot if you were a doctor or nurse, for being both vulnerable and important, but not if you were a philosopher or lawyer (clearly not as important in this context!). In other words, if you did not belong to a group that was particularly vulnerable to the virus, your chances of surviving due to vaccination depended on how beneficial you were for society in terms of fighting the pandemic, which sounds like a revival of the older triage rules. There is more to it than being beneficial for society, however. Maybe it is also important whether your specific personal characteristics put you in a position that is regarded more highly by society. Is it not, therefore, the case that women and children should definitely come first?

2.5 Women and Children First!?

Remember that the recent Italian guidelines for intensive care doctors ascribed great importance to treating those who can gain more life years. This, Professor Skeptico points out, mirrors the strongest preference the Moral Machine researchers found among their more than two million participants worldwide: People seem to have a strong preference to save younger over older lives. If a young person and an old person have exactly the same medical condition, it is the younger person who is statistically expected to gain more life years simply because they are younger.

While our equality-minded Professor Principia regards this as yet another group-based discrimination that is clearly morally wrong, Skeptico insists that age as a category is different from other group-based categorizations (Černý 2022). Of course, everyone belongs to a certain age group, just as they belong to a certain religious (or non-religious) group, have a certain skin tone, and a certain gender. The difference, however, is that we all start young and grow older (if we are lucky). In contrast, we do not usually start with one religion, skin tone, or gender and then change them every decade or so (although some do).

Moreover, those who are old now profited from the greater protection of the young when they were young. Those who are young now benefit from the same protection but will be less protected when they are older. Once the largest part of their lives is behind them, it is just fair to prioritize the young, who still have so much in front of them. *Principia* quickly clarifies that since self-driving vehicles are not yet out on the streets in significant numbers, this argument will only apply in the future when the first generation that was prioritized by self-driving vehicles as children grow older.

True, Professor *Skeptico* says, but goes on to explain that the idea of prioritizing the young in emergency situations is not new. In May 1840, a fire broke out on the *Poland*, an American packet boat on its way from New York to Le Havre, France. As one of the passengers reported, when people realized that they had to get to the lifeboats,

> it was immediately remarked by one of our French passengers, and responded to by others—'Let us take care of the women and children first.'
>
> *(Howland 1840)*

This is the first reported mention of the 'women-and-children-first' rule that was later used in many emergencies at sea, including the dramatic sinking of the *Titanic* in 1912.

However, Professor Principia cautions that this rule, despite its ubiquity in public culture, was never part of maritime law. Also, it is perhaps not as convincing as it looks at first glance because it applies to children and women alike, and it is not clear why. You might well believe that there are many good reasons to prioritize children being rescued. After all, they are innocent in the sense that they typically did not decide to make the boat trip in the first place, they still have the longest time of their lives ahead of them, and they might be unable to help themselves in emergency situations, etc. But what exactly speaks for prioritizing women? At least today they are more likely to have had their say in arranging the trip and they are not younger than men per se. So is there any reason that is not plain patronization or that does not rest on gender stereotypes? Do you believe that the children and the women should be treated preferentially because they cannot help themselves, or that the children need their mothers more than their fathers?

2.6 Are These the Wrong Questions to Ask?

While both professors are established academics seeking to put forward convincing arguments, it is important at this stage that we do not come to hasty conclusions by following the arguments of only one of them, or by relying on empirical findings too readily. While the Moral Machine Experiment certainly has the largest data set so far, it is less certain whether we can identify the moral preferences of the general public in the way that the researchers behind the experiment thought.

Remember that we started the discussion of the moral relevance of personal characteristics with the findings of the Moral Machine Experiment. Based on an enormous data set, the conclusion drawn was that around the globe, people prefer to save younger over older lives, people with a higher social status (or those who benefit society more), and women before men. These preferences are clearly not in line with the German Act on Autonomous Driving, or with the recommendation of the Ethics Commission. Both prohibit any kind of qualification according to personal characteristics. While Professor *Skeptico* thinks that a discrepancy between what people think and what the law says is deeply unsettling when it comes to fundamental moral issues that potentially affect the everyday lives of almost everybody, Professor Principia suggests a smooth way to resolve the tension between the law and people's opinions. In fact, she has reasons to believe that there is no tension at all.

Terrified by the results of the Moral Machine Experiment, Principia dug deeper into the debate that emerged from it and found a particularly intriguing set of (related) criticisms of the Moral Machine Experiment. The main criticism, she explains, is straightforward: Perhaps some of the results of the Moral Machine Experiment are prompted by a flaw in the design of the study. Remember that the researchers presented participants with situations with two exhaustive alternatives: either the car kills person A or it kills person B. Based on specific information about A (a doctor!) and B (a criminal!), the participants had to decide who should be killed by the car.

What that either/or design lacks, two psychologists, Yochanan Bigman and Kurt Gray, brilliantly suspected, is a third option, namely one that allows the participants to treat the potential victims equally (Bigman and Gray 2020). If they were right, the results found in the Moral Machine Experiment would represent forced decisions. That is, people had to decide due to the research design, not because this reflects what they actually think. It is a bit like being asked if you prefer red wine or white wine even though you actually prefer beer. Being psychologists, Bigman and Gray naturally ran their own experiments to test whether participants' responses would be different when a third option was added: either the car kills person A, or it kills person B, *or it treats the lives of person A and B equally*.

As Principia explains, this equality option means that the car simply ignores (or fails to detect) the personal features of A and B. In other words, the car just would not (be allowed to) know who a doctor or a criminal, a child or a granny is; it would simply recognize that there is a human being and treat everyone equally. Note that this is something that human drivers cannot do. They are not able to *not* recognize (intentionally) if the person before them is a child or a granny, which means that there is always a certain likelihood that they will take such characteristics into account.

Indeed, the experiments show that many participants seem to prefer such an equality option over killing either A or B. For example, they found that with that additional option, the preference to save children over the elderly dropped from 96.1% to 38.8%, whereas 61.1% preferred the equality option to treat children and the elderly alike. Similarly, the preference to save those with high social status drops from 79.7% to just 12.7%. The vast majority (84.7%) want to treat potential victims with high and low social status equally.

Although these findings are based on responses from fewer than 1,000 participants (compared to the 2.3 million participants in the Moral Machine Experiment), they were significant enough to merit publication in the journal *Nature*. As Principia triumphantly announces, when you ask the wrong questions, no sample size will give you the relevant insight.

Skeptico concedes that it really does seem as though what the Moral Machine Experiment investigated is something like forced preferences between killing A or killing B although many participants might have preferred another option that was just not on the table. So maybe there is no tension between the law and the general public's opinions. They want what the law says, namely to prohibit any programming of self-driving cars that qualifies potential victims according to their personal characteristics.

For now, Skeptico is licking his wounds. But he already senses that the findings of Bigman and Grey might have implications that Principia will not like. They have to do with another moral dilemma, which has been the center of much academic discussion by lawyers, philosophers, and psychologists for many decades, and which seems to be highly relevant for self-driving vehicles.

2.7 The Numbers Problem

There is another kind of dilemma situation that is likely to occur with self-driving vehicles, namely those in which one person can be sacrificed in order to save a larger group of people. Imagine a situation in which a self-driving car can either run into one pedestrian, which would kill them, or into five pedestrians, which would do the same. Should the car be programmed to kill one person or five? Should the numbers count?[5]

The German Act on Autonomous Driving is silent on such dilemma situations. Remember that this law is based on the recommendations of an Ethics Commission. As the Commission's report shows, its members—esteemed experts in various fields relevant to the topic, chaired by a former justice of the German Federal Constitutional Court—could not reach a consensus about what to recommend for such situations. Most importantly, they could not agree on the (im-)morality of a program saving as many lives as possible in such cases. Trade-offs between human lives are, thus, currently neither explicitly prohibited nor explicitly allowed by the law (Hilgendorf 2021).

This is unfortunate. Granted, these dilemma situations pose moral problems that are very hard to solve. But as we explained in Chapter 1, there is simply no option *not* to solve them. Car manufacturers have to program their cars to deal with such dilemmas *somehow*. Whatever the program is set to do in these situations involves a moral decision with fatal consequences. No matter whether the algorithm says 'do nothing' and stay on course, or 'save the greatest number,' say, each option is an answer to the moral problem. And because each option will result in the killing of some (and the saving of others), it had better be well justified. In this section, we will explore possible solutions to such dilemma situations and how they can be justified.

Consider a fictitious scenario that is somewhat similar to Kohler's 1915 Car Driver case, namely the Cargo Wagon case, invented by another German criminal lawyer, Hans Welzel, in 1951:

> A cargo wagon has broken loose on a steep mountain track and is rushing down into the valley at full speed towards a small station where a passenger train is currently standing. If the wagon were to continue on its present track, it would hit the passenger train and kill a large number of people. A railway official, seeing the disaster coming, pulls over the switch at the last minute, which directs the cargo wagon onto the only side track, where some workers are unloading a wagon. As the official foresaw, the collision kills three workers.
>
> *(Welzel 1951, 51, translation by the authors)*

How would you judge the railway official? Similar to Kohler, Welzel argued that the railroad official is not *culpable* for having redirected the cargo wagon, even though this killed three people, because doing so is regarded as excusable. Note, however, that Kohler's and Welzel's cases focus on the action of an individual and on the question of whether this individual should be punished or not. Remember that in Kohler's case, the question was whether the driver would be culpable for killing person *A*, *B*, or *C*.

When we think about how to regulate dilemma situations with self-driving vehicles, this perspective of criminal law is not very informative. After all, Kohler's and Welzel's point is that the agents are not culpable, no matter which option they choose. In fact, this is merely an acknowledgment of the moral dilemma the agents are in, and not blaming them for choosing between two evils. From this perspective, those who find themselves in dilemma situations are only to blame if they try to take advantage (as when someone abuses such a situation by intentionally killing their nasty mother-in-law, remember?). But the situation with self-driving vehicles is precisely about we as a society being able to decide who should be killed and who should be saved *in advance*. The primarily relevant question is how the cars should be programmed to act. Who is legally responsible—and maybe even criminally culpable—for a fatal accident with self-driving cars is certainly a very important question, but one that hinges on the answer to the first question. Only once we know what the car should have done can we say who is responsible for what (if the car did not do what the law required, say).

2.8 Intending vs. Foreseeing: The First Trolley Case

Welzel's Cargo Wagon case might remind you of a philosophical thought experiment that is well-known even outside the academic ivory tower, the so-called Trolley cases. The first Trolley case was invented by the Oxford philosopher Philippa Foot, née *Bosanquet*. On her father's side, she came from a wealthy English family with a long pedigree; the other part of her family was less posh, her mother was the daughter of U.S. President Grover Cleveland (to this day the only presidential child born in the White House) (Edmonds 2013, chap. 3). Here is her case (we'll call it the Driver case):

> [Someone] is the driver of a runaway tram which he can only steer from one narrow track on to another; five men are working on one track and one man on the other; anyone on the track he enters is bound to be killed.
>
> *(Foot 2003, 23)*

Replace the tram with a car and you have a version of Kohler's case, but one in which the driver is bound to kill either one or five innocent people. Unlike Kohler's, *Foot*'s fictitious scenario—captured in just one unassuming sentence—had an enormous impact on philosophical debates in the following decades. The article containing this thought experiment was published in 1967 in the *Oxford Review*; it is not about trolleys (or trams, in *Foot*'s English parlance); its title is 'The Problem of Abortion and the Doctrine of the Double Effect' (Foot 2003).

The doctrine of double effect was most prominently developed by Thomas Aquinas, a 13th-century Dominican friar whose philosophical writings resonate to this day. After his death, he was made a saint—an honor that few philosophers since could reasonably expect. The doctrine of double effect focuses on the distinction between intending and foreseeing. For example, intentional killing can never be justified. But consider cases of self-defense. The doctrine holds that killing can be permissible under certain circumstances, namely if the actual intention of killing someone is self-preservation, the death of the person being foreseen by the actor as a consequence of the intentional act (at least as long as there is no harmless way to bring about the intended result, and the harm is not disproportionate to the intended aim). It is called the doctrine of double effect because it builds on two effects some actions can have—one strictly intended, the other foreseen but not intended. In Catholic theology, this doctrine is used to justify the prohibition of abortion. Viewed from the perspective of the doctrine of double effect, an abortion clearly involves the intentional killing of a fetus and is, thus, forbidden. But the doctrine allows for exceptions, for example, if a woman's life is at stake and a certain procedure is needed to save her. If that procedure—intended to save the woman—inevitably results in the death of the fetus, this bad consequence is regarded as merely foreseen, and the procedure is, thus, justified.

This distinction between the strictly intended good consequences and the merely foreseen bad consequences of one and the same act has been very influential in legal and moral discourse. It remains an important element in many jurisdictions as well as in medical practice, for example in the distinction between the permitted administering of a drug that is intended to alleviate a patient's pain but is foreseen to end their life a bit earlier on the one hand and the prohibited administering of a drug that is intended to hasten the patient's death on the other. Even though the result is the same—the patient dies today instead of tomorrow, say—the intention of alleviating pain vs. shortening life makes the difference, at least according to the doctrine of double effect.

What does this have to do with the Trolley cases and with self-driving cars, you might ask. Well, the answer has to do with yet another case. In order to illustrate the plausibility of the doctrine of double effect, *Foot* designed the Driver case in contrast to the Judge case:

> Suppose that a judge or magistrate is faced with rioters demanding that a culprit be found for a certain crime and threatening otherwise to take their own bloody revenge on a particular section of the community. The real culprit being unknown, the judge sees himself as able to prevent the bloodshed only by framing some innocent person and having him executed.
>
> *(Foot 2003, 23)*

As Foot explains, the 'question is why we should say, without hesitation, that the driver should steer for the less occupied track, while most of us would be appalled at the idea that the innocent man could be framed.' Her initial answer is that the doctrine of double effect rightly points out that 'it is one thing to steer towards someone foreseeing that you will kill him and another to aim at his death as part of your plan.' In other words, the judge needs the innocent man to die. His death is part of the judge's plan to save more people who would otherwise die in the riot. The tram driver, in contrast, would be relieved to see the person on the side track jump off the track before the train hits him. It is not his plan to kill the worker; his death is merely the foreseen (bad) effect of his (good) intention to save the five workers.

Foot dismisses the doctrine of double effect because in her view it conflates direct intention and mere foreseeing on the one hand with actively doing something and merely allowing it to happen on the other. Her idea was to draw on the distinction between positive and negative rights instead. Negative rights and their corresponding duties (i.e., what we owe to others in terms of non-interference) are stronger, she argues, than positive rights and their corresponding duties (i.e., what we owe to others in the form of aid).[6] Foot explains that the judge has the negative duty of not inflicting injury and the positive duty of bringing aid. As negative duties are stronger, the judge must not sacrifice the innocent man. Can you already see where this is going and how the Judge case helps find a solution to the Driver case?

In contrast to the judge, the tram driver faces a conflict between two negative duties, namely the duty of not killing the five men and the duty of not killing the one. Since the driver is not able to fulfill both negative duties, he should save as many people as he can, or so *Foot* argues. And if we replace the tram with a car, this would also apply to a car driver. Isn't that great?, says Skeptico. *Foot*'s solution gives us an answer to the moral question as to whether the driver should save as many lives as possible. Of course, he should, at least if the dilemma is one in which two negative duties conflict. But wait, Principia replies, I thought we were interested in self-driving vehicles, not in drivers making fateful decisions. Moreover, wasn't Foot's reasoning criticized in the ensuing debate?

2.9 Serious Trouble for 'Mrs. *Foot*'

Of course, it was. Most prominently so by the American philosopher Judith Jarvis Thomson. Thomson developed many Trolley cases, among them a version of the Driver case that is now often regarded as *the* Trolley case, namely the Bystander case. In the Bystander case, it is not the driver who can change the course of the trolley but a random bystander who happens to be standing near a switch. If he throws the switch, the trolley will be redirected onto the side track, killing the one worker.[7]

This case brings us one step closer to self-driving cars. After all, now the trolley has no driver and its course is determined by an external actor, the bystander. In that sense, the bystander is at least somewhat similar to a self-driving car's autopilot. It is, thus, not surprising that scenarios like the following are routinely discussed in relation to the ethics of self-driving vehicles:

> Six men are crossing the road. A self-driving car is approaching fast and cannot stop before it gets to the pedestrians. If it stays on course, it will kill five of the men. If it swerves, it will kill the other man. It just cannot save all of them. Should the car stay on track or change course?

Figure 2.2 shows one image from the Moral Machine Experiment that tests exactly such a scenario.

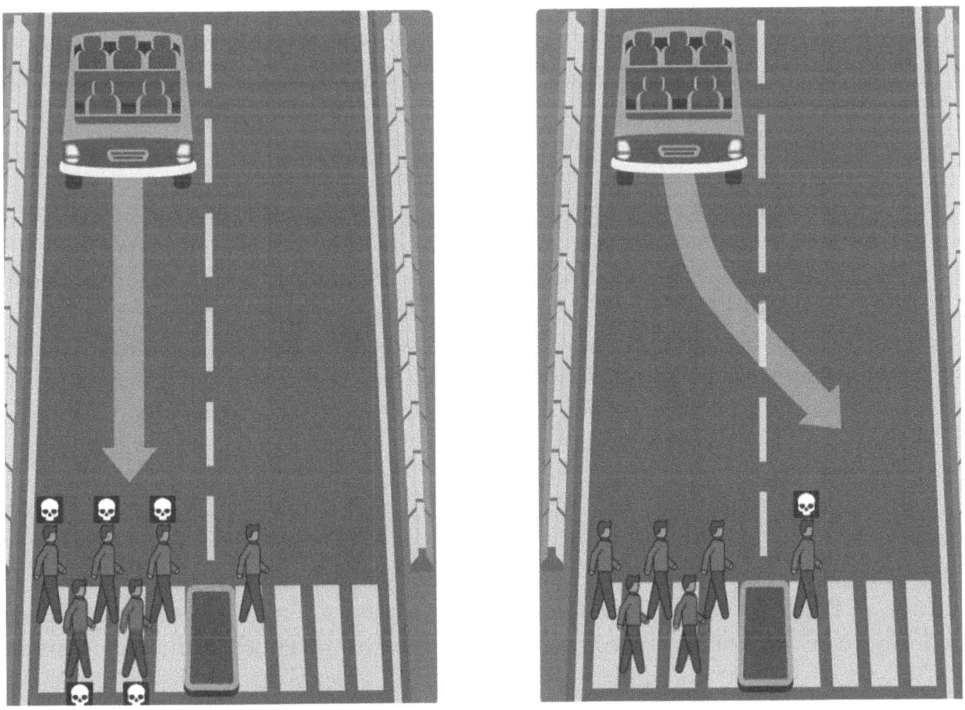

FIGURE 2.2 An image from the Moral Machine Experiment asking participants to decide between killing one person and killing five people.

The Bystander case is very similar to Welzel's Cargo Wagon case. In both cases a bystander can redirect a train in order to save a larger number of people, thereby sacrificing a smaller number. As far as we know, neither Foot nor Thomson was familiar with Welzel's case.[8]

In any case, it was Foot's Driver case and Thomson's diverse variations on it, which really did kickstart the famous trolley debate. Moreover, the emerging debate had little to do with the question as to whether the bystander/railroad official is to be regarded as criminally culpable. If there is one thing to take away from reading this book—written by an Austrian lawyer and a German philosopher—it is that those who were familiar with Welzel's case had pure philosophical gold in their hands, at least in between the publication of *Foot's* Driver case (1967) and Thomson's Bystander case (1983).

But why was it philosophical gold? Why was it an important development to introduce the Bystander case? Are the two cases—Driver and Bystander—not morally similar? Well, do you find it morally acceptable for the bystander to redirect the trolley? Thomson believes that most people would find that acceptable, just as it would be acceptable for the driver in Foot's case. There is only one problem: The bystander violates a negative duty (to not kill one person) in order to fulfill a positive duty (to help the five people who would die if he did nothing), which contradicts Foot's suggestion that the relevant moral difference between cases such as Driver and Bystander lies in the distinction between positive and negative rights and duties.[9] This, Thomson said, 'is serious trouble for Mrs. Foot's thesis' (Thomson 1985).

2.10 The Trolley Problem

However, the question that was discussed in the following decades was not what distinguishes cases like Thomson's Bystander case from Foot's Driver case. Instead, the question revolved around the difference between the Bystander case and yet another variation of the Driver case, namely Thomson's Footbridge case:

> Consider a case [...] in which you are standing on a footbridge over the trolley track. You can see a trolley hurtling down the track, out of control. You turn around to see where the trolley is headed, and there are five workmen on the track where it exits from under the footbridge. What to do? Being an expert on trolleys, you know of one certain way to stop an out-of-control trolley: Drop a really heavy weight in its path. But where to find one? It just so happens that standing next to you on the footbridge is a fat man, a really fat man. He is leaning over the railing, watching the trolley; all you have to do is to give him a little shove, and over the railing he will go, onto the track in the path of the trolley. Would it be permissible for you to do this?
>
> *(Thomson 1985, 1409)*

After presenting the scenario, Thomson adds: 'Everybody to whom I have put this case says it would not be. But why?' She assumes that the fat man must not be pushed off the bridge in the Footbridge case. This judgment, too, cannot be explained by Foot's distinction between positive and negative rights and duties—at least not with an explanation that is consistent with the moral judgment in the Bystander case. In both cases, an innocent person has to be actively killed in order to save five people. If this is morally permissible in the Bystander case but not in the Footbridge case, then we face what Thomson has coined the 'trolley problem, in honour of Mrs. Foot's example [the Driver case]' (Thomson 1976). Here is *Thomson*'s formulation of the trolley problem:

> Why is it that the bystander may turn his trolley, though [one must not push the fat man off the bridge]? Since *I* find it particularly puzzling that the bystander may turn his trolley, I am inclined to call this The Trolley Problem.
>
> *(Thomson 1985, 1401)*[10]

You might think that with the Footbridge case we are finally losing our minds. What on earth does this scenario have to do with self-driving cars? Well, consider the Microcar case:

> A self-driving car without passengers is driving down the road. As it approaches a crossing where the traffic light is red, its brakes fail. If the car continues on its course, it will kill a group of five people. The only way to stop this from happening is to direct a self-driving microcar, a tiny car with only two seats (such as the Renault Twizy), with a fat man on board, which is approaching the crossing from the other direction, into the path of the out-of-control car before it can hit the group of five. Unfortunately, this microcar has one passenger that would be killed if it collides with the out-of-control car. Without the passenger the microcar would not be heavy enough to stop the other car.[11]

This scenario is, admittedly, somewhat far-fetched but it is not totally unrealistic. After all, once self-driving cars are on the road in significant numbers, they will be connected. The connection between them is one of the main tools to make them drive more safely than human drivers. Through its connection to other vehicles, each car will "know" the traffic situation everywhere along its route, far beyond what can be "seen" with its sensors, let alone with human eyes. This will enable such interconnected cars to anticipate dangerous situations much better than human drivers can.

Summing up, the trolley problem is how the difference can be explained between the moral judgments in the Bystander case and the Footbridge case. That problem is, of course, not limited to trolleys. It concerns a certain structure that can also be found in a host of other cases, including dilemmas with self-driving vehicles.[12]

There have been numerous attempts to solve the trolley problem, i.e., to suggest moral principles that can convincingly explain the different moral judgments. However, none of these has convinced many philosophers (Bruers and Braeckman 2014). It is not by accident that one of the most inventive "trolleyologists", the American philosopher Frances Kamm, who has spent decades trying out different possible solutions, entitled her most recent book on the topic *The Trolley Problem Mysteries* (Kamm 2016).

2.11 How the Trolley Problem Might Help Us Regulate Self-Driving Vehicles

Nevertheless, the trolley problem might provide us with a solution to some of the moral problems with self-driving vehicles. The idea of how this might work is relatively simple. When philosophers seek a solution to the trolley problem, they suggest moral principles that can convincingly explain the different moral judgments in the Bystander and Footbridge cases.

Remember, these judgments are that the bystander may hit the switch (killing the one person on the side track) but that the person on the bridge may not push the fat man onto the tracks (killing him). The whole philosophical debate that evolved around Trolley cases assumes that these two judgments are correct, namely that it is morally permissible to hit the switch in Bystander but that it is morally forbidden to push the fat man in Footbridge. The various solutions that have been offered differ in how they account for that difference but all of them accept the moral judgments in the Bystander and Footbridge cases.

These judgments are not only assumed to be right for the sake of argument, though. Responses to the Trolley cases have been investigated empirically in many studies. In fact, there are so many studies that use scenarios modeled on the Trolley cases that they are said to constitute moral psychology's '*lingua franca*' for studies of moral decision-making (Cushman 2013). According to such studies, approx. 90% of laypeople find it permissible to hit the switch in the Bystander case while only approx. 10% find it morally OK to push the fat man off the bridge in the Footbridge case (Hauser et al. 2007). Among professional philosophers, approx. 63% find it permissible to redirect the trolley in the Bystander case while only about 13% find redirecting it impermissible (Bourget and Chalmers 2020). Approx. 22% in turn find it OK to push the fat man while approx. 56% are against pushing him. These findings with laypeople and experts alike thus show that the trolley problem addresses a real moral problem that has defined recent moral philosophy.[13]

Do you already see how this can help us solve moral dilemmas with self-driving cars? Take the Bystander case. In the debate about the trolley problem, it is assumed to be permissible for the bystander to hit the switch. Now take the case of a self-driving car that will run into a group of five (killing all of them) if it is programmed to stay on track or will kill one of them if it is programmed to swerve. Assuming that this scenario is morally similar to the Bystander case, programming self-driving cars to let the car swerve (killing one) in order to save a larger group of people is also permissible. If the two cases are morally similar, they warrant the same moral judgment.

The same thought applies to the Footbridge case and the Microcar case. Insofar as they are morally similar, programming self-driving cars to use one car (with a passenger) to prevent another car (without passengers) from running into and killing a larger group of people, leading to the passenger's death, is prohibited because it is also prohibited to push the fat man in the Footbridge case.

There is also some empirical evidence that speaks for the analogy between Trolley cases and cases with self-driving cars. We discussed some of the findings of the Moral Machine Experiment above but only insofar as they concerned the personal characteristics of the potential victims (such as social status, age, or gender). The experiment also revealed insights into people's preferences regarding the question we are facing in this section, namely trade-offs between human lives. The researchers found that the tendency to save as many lives as possible is one of the most dominant moral preferences to be found worldwide. In other words, most people would say that cars should be programmed to swerve in order to save the five people crossing the road, even if this kills the other person—and people show that preference over a large range of similar scenarios that are roughly similar to the Bystander case.

But did we not conclude that the Moral Machine Experiment actually forced certain preferences through its either/or design (either the car kills person A or person B) although many participants might have preferred another alternative option that just was not on the table? Now is the time for Skeptico's rebuttal. He sensed that Bigman and Grey's findings would have implications that Principia would not like. Remember that Bigman and Grey found that people wanted a third option, an equality option, according to which everyone counts the same, no matter whether they are a doctor or criminal, say.

But what does this mean when it comes to trade-offs between different numbers of human lives? Well, when everyone counts the same, one person counts less than two, two less than three, and so on. This is indeed what Bigman and Grey found: 81.6% of their participants preferred to save more lives and only 17.9% wanted to treat different groups of potential victims equally in the sense that group size should not make a difference (Bigman and Gray 2020).[14] But again, this only holds for Bystander-like scenarios. As far as we know, scenarios like the Microcar case that are similar to the Footbridge case have not yet been investigated empirically.

2.12 Self-Driving Cars Are Not Like Trolleys!?

In the previous section, we laid out the case for finding a solution to dilemmas with self-driving cars by building an analogy to the original philosophical Trolley cases. There are multiple ways of criticizing this approach. Perhaps the most obvious one is to argue that the analogy between trolleys and self-driving cars does not hold. For instance, trolleys run on tracks, but cars do not. If you divert a trolley, you know exactly where it will go because it will follow the track. If a self-driving car is programmed to swerve, this can result in many different trajectories depending on how much it swerves. This means that it is much more uncertain where exactly a car will go. It is very hard to say if it will hit this person or another one, or if it can manage to avoid a deadly crash altogether.

For philosophical discussions about the trolley problem, it is assumed that the outcomes are fixed and known with certainty; in other words, full knowledge exists about the outcomes of each possible action or inaction. Take the Footbridge case. When describing the scenario, Thomson says that '[b]eing an expert on trolleys, you know of one certain way to stop an out-of-control trolley: Drop a really heavy weight in its path' and that '[i]t just so happens that standing next to you on the footbridge is a fat man, a really fat man' (Thomson 1985, 1409). There are many things that she does not say there but must be assumed for the dilemma to make sense. For instance, she must assume that it is clear and known that the five on the track will die if the agent does not intervene; that pushing the fat man will definitely stop the trolley (and, thus, save the five); that this action will kill the fat man; that the agent is able to topple the fat man over the railings on the bridge; and, importantly, that there is no other heavy object (the agent, say) that could stop the trolley. What do you reckon? How often does such a situation occur in real life? It is probably fair to say: never!

This may come as a surprise to some readers, but even philosophers know that this is not how things work in real life. As Foot said right after introducing the mother of all Trolley cases, the Driver case: 'In real life the certainties postulated by philosophers hardly ever exist' (Foot 2003, 31). So why do so many philosophers spend decades thinking about totally unrealistic cases?

It is the same reason why scientists study the natural world in a lab. Because there the environment can be designed in such a way that the relevant features can be investigated. It is almost impossible to do much advanced science in the natural environment. There are just too many factors that can influence the thing under investigation. In a lab, you can design the environment such that the most relevant factors are isolated. And by manipulating these factors individually, you can observe what they do. This is precisely what philosophers do, only they do not need a lab. This is why philosophers call the Trolley cases and other such scenarios thought experiments. To borrow a phrase from James R. Brown, they are executed in the 'laboratory of the mind' (Brown 1991).

In this laboratory, philosophers design the environment such that the issue under investigation—here, the trolley problem—can be isolated from the infinite variety of other factors. As a moral issue, the trolley problem is interesting irrespective of all of the other factors that potentially determine what happens in real-life accident situations. Maybe the workers manage to jump off the rails just in time; maybe the fat man only gets injured but survives, or even he is not heavy enough to stop the trolley, etc. None of that matters for the trolley problem because it addresses a fundamental moral issue that has to do with the structure of each individual case. Why does it seem right to divert the trolley in the Bystander case but not to push the fat man in the Footbridge case? Similarly, why does it seem right to program self-driving vehicles to swerve if this saves more lives but not to use a second car to obstruct an out-of-control car in the Microcar case although this would also save more lives?

Having isolated a moral issue in this way, it is much easier to think about the one issue at hand without being distracted by the huge number of other factors present in real life. It is often said that philosophers should use realistic, or at least more realistic, scenarios for their thought experiments. This sentiment is understandable, but it misses the crucial point that they are deliberately simplifying things in order to make difficult issues in real life accessible. Trust us, if it were easier to think through difficult moral problems in complex real-life situations, philosophers would be thrilled to do just that. But it just is not. It is more likely that we lose sight of the precise moral issue if we try to think through problems that are as difficult as the trolley problem in a real-life setting, with all the potentially relevant factors in place, with all the known and unknown unknowns.

2.13 Self-Driving Vehicles Are Going to Cause Dilemmas, Unfortunately

In a way, the discussion of dilemmas with self-driving vehicles is more realistic than the traditional trolley cases.[15] As we argued in Chapter 1, there certainly will be accidents with self-driving cars. And it is reasonable to expect that a tiny number of these will be dilemmatic in the sense that the car can only avoid one of two evils. It is equally reasonable to expect that there will be even fewer situations in which the car recognizes with certainty that, after brake failure, say, it only has two options, one of which will result in the killing of three people and the other in the death of one person. Autopilots are designed to ascribe probabilities to each possibility. So they might identify an object in front of them as a human being, but they will never know for certain if it really is a person or maybe just a statue. They will also never know for certain that, if it is a person and if they crash into them, this person will inevitably be killed. Maybe they will just be seriously injured. Is this not a good reason to stop focusing on dilemmas involving self-driving vehicles and instead focus on avoiding accidents in the first place?

Not necessarily. It is of the utmost importance to design self-driving vehicles such that they cause as few accidents as possible. This does not imply, however, that they will never face dilemma situations. Certainly, these situations will be rare. But it is by no means unrealistic to assume that a situation such as the following can happen (Evans and Furey 2022):

> A self-driving car's autopilot identifies two people (it is 95% certain, say, that these objects really are humans) in its way who it will most likely hit (with 90% certainty, say), which will maybe kill them (70%, say, because the car is driving fast); searching in milliseconds for a safer trajectory, the autopilot realizes that there is only one other option, namely to swerve to the right, where it identifies a third person (it is, again, 95% certain that this really is a human being), who it will most likely hit (again with 90% certainty), which will perhaps kill them (again 70%).

This description is much more realistic than the scenarios with self-driving cars that we have discussed so far. But does acknowledging limited certainties make any difference from a moral point of view? Think about it: what should the car be programmed to do in such a scenario that is definitely more realistic? The moral issue is not what the chances are that the object really is a human being or that the crash will kill a person. For, clearly, if it is highly unlikely that the crash will kill a human being, then the case is not at all like the dilemmas we are interested in. But what if it is very likely that the crash will kill one, two, or three people? Well, then you are back to discussing the trolley problem.

2.14 Always Save the Many? Yes, Says *Bentham*

As we have seen above, most philosophers and laypeople agree with the judgments assumed for the traditional Trolley cases. But some philosophers do not. There are basically two philosophical camps that regard the trolley problem as being misguided. One of these camps holds that the fat man should also be pushed off the bridge because this would result in more lives being saved. The other camp holds that it is prohibited to push the fat man in the Footbridge case and to divert the trolley in the Bystander case because both acts would violate the human dignity of the one person being killed. Both camps have famous (dead!) philosophers who can provide them with good arguments. After considering these two positions, we will get to know a more recent moral theory called contractualism.

Those who focus on the number of victims stand in the tradition of the English philosopher Jeremy Bentham. Bentham died in 1832, but his skeleton padded out with hay and dressed in his clothes with a wax head instead of his mummified one can still be seen at University College London. He developed a moral doctrine according to which what counts morally is how much pleasure an action produces and how much pain it avoids. It is all about the consequences of an action. The catchphrase of Bentham's so-called utilitarianism is 'the greatest happiness of the greatest number' (Bentham 1977, 393). As everyone counts the same according to this standard, utilitarianism is deeply egalitarian. It helped pave the way for a whole range of progressive societal reforms. Since all that matters morally is pleasure and pain, Bentham thought that we should consider both human and animal feelings. Similarly, if love and sex bring pleasure, it does not matter if it is between a man and a woman, or between two men. *Bentham* did not stop there, why not between man or woman and animal (Edmonds 2013, chap. 8)?

So for Bentham, all that matters morally is the consequences of an action, how much pleasure it produces and how much pain it avoids. Naturally, then, he was not interested in subtle differences such as the one between intending and foreseeing, between killing someone and letting them die, between positive and negative rights, and so on. He just did not believe that any of these are morally relevant. In fact, his theory does not even differentiate between different qualities of pleasures. Simple animalistic joys are just as good as highly cultivated pleasures. This was criticized by many commentators. They argued that Bentham degrades humans to the status of animals. It fell to John Stuart Mill, the philosopher who invented the harm principle, to defend utilitarianism by developing a version of it that includes higher pleasures. In Mill's words, '[i]t is better to be a human being dissatisfied than a pig satisfied; better to be Socrates dissatisfied than a fool satisfied. And if the fool, or the pig, are of a different opinion, it is because they only know their own side of the question' (Mill 1969, 212).

The irony of the story is that Bentham was Mill's spiritual foster father, or maybe foster grandfather—Mill was born in 1806 when *Bentham* was already almost 60 years old. Bentham not only provided financial help to his friend James Mill, John Stuart's father, when the family struggled. The two also decided to run an experiment on John Stuart. Basically, they deprived him of a normal childhood and educated him in accordance with their rationalistic ideal very early on. As a young child, John Stuart read Plato and *Aristotle* in Greek; at the age of six he wrote a history of Rome; for relaxation, the boy read Plutarch and Hume's History of Great Britain. When he was eight, John Stuart was made responsible for teaching his younger siblings in Latin. When the siblings made mistakes in their translations, *John Stuart* was punished with food deprivation (Reeves 2008, chap. 1).

Isaiah Berlin describes this educational experiment of Bentham and James Mill as 'an appalling success' (Berlin 1969, 175). After all, John Stuart became a remarkable intellectual and politician. Among other things, he wrote—heavily influenced by his wife Harriet Taylor Mill—influential books on utilitarianism, liberty, the economy, and, last but not least, the subjection of women. A few years after the premature death of his wife in 1858, Mill even served as a member of parliament with the aim of achieving women's suffrage. He did not succeed, of course. But his work helped pave the way for the eventual enfranchisement of women (at least those over 30) to vote in the United Kingdom in 1918.

For utilitarians like Bentham and Mill, the only thing that counts is the greatest happiness for the greatest number: The trolley problem would not be a moral problem. In the Bystander case, either five people or one can be killed. So clearly the lone individual should be killed. In the Footbridge case, either five people or one can also be killed. So here again, clearly the lone individual should be killed. For Bentham, the cases have the same moral structure and, thus, warrant the same solution. He would certainly program self-driving vehicles such that they always save as many lives as possible.

2.15 Always Save the Many? No, Says *Kant*

Another approach that disagrees with the judgments assumed for the traditional Trolley cases emphasizes a notion that traditional utilitarians such as Bentham have problems making sense of, namely individual rights. When you have a right, you are like 'a small scale sovereign' (Hart 1982, 183). Rights have the function of protecting and fostering individual autonomy. They are regarded as personalized claims that are untradeable across individuals, or that trump other considerations such as public interests. Arguably the most fundamental right is human dignity. Some even think that it is an absolute right, i.e., one that can never be limited. This emphasis on rights and their corresponding duties gives this theory its name: deontology.

The idea of a special dignity that comes with belonging to the human race, and that individuals have irrespective of their social standing, abilities, or other personal characteristics, has a long tradition. Today it is to be found in almost all European constitutions—not in the United Kingdom, though. Modern uses of human dignity mostly stand in the tradition of the Prussian philosopher Immanuel Kant, whose so-called categorical imperative is one of the most important foundations of modern ethics. This imperative is a categorical one because it is an absolute moral requirement. It tells you what to do—or rather, what not to do—in all situations and irrespective of your desires, full stop. A famous example is that you shall not lie, even in the most extreme circumstances. Imagine that you are hiding a perfectly innocent person in your wardrobe who is being hunted down by an angry killer. Even then, Kant says, if the killer asks you if you know where his future victim is hiding, you should tell the truth. After all, you shall not lie, full stop.[16]

Here is one of Kant's formulations of the categorical imperative: 'So act that you use the humanity, whether in your own person or in the person of any other, always at the same time as an end, never merely as a means' (Kant 1996, 80). Consider our Footbridge case, for instance: The fat man is used merely as a means because you have to push him off the bridge. You need him to stop the approaching trolley.

The German Ethics Commission has the *Kantian* categorical imperative in mind when it says in the context of self-driving vehicles that 'the sacrifice of innocent people in favour of other potential victims is impermissible, because the innocent parties would be degraded to mere instrument and deprived of the quality as a subject' (Ethics Commission 2017, 18). You shall not lie, and you shall not push the fat man off the bridge, you shall not make a person the victim of a self-driving car accident just for the sake of saving more people. And the report continues:

> Even in an emergency, human lives must not be "offset" against each other. According to this position, the individual is to be regarded as "sacrosanct". No obligations of solidarity must be imposed on individuals requiring them to sacrifice themselves for others, even if this is the only way to save other people.
>
> *(Ethics Commission 2017, 18)*

This 'sanctity' of the individual is at the core of the idea of human dignity. It blocks all attempts to engage in trade-offs between human lives. The implications for self-driving vehicles are clear: They must not be programmed to always save the greatest number. Thus, in the Microcar case (similar to the Footbridge case), the second car must not be used to obstruct the path of the out-of-control car. This is still in line with the judgments assumed in the trolley problem. But in situations similar to the Bystander case, the *Kantian* idea of human dignity is often taken to imply that the car must not swerve but crash into the five people, even if this kills them all (Ethics Commission 2017; Schmidt 2022).[17] However, others argue that Kant can be interpreted such that the bystander does not 'use' the one person on the side track because this does not contribute to the saving of the five persons on the other track. The trolley would also be diverted to the side track if the one person were not on that track (Kleingeld 2020; Guyer 2014; Kerstein 2013).

Remember Thomson, the philosopher who first formulated the trolley problem and imagined some of the most iconic Trolley cases? Decades after she did that, she changed her mind and settled on a position similar to that of the German Ethics Commission (Thomson 2008). Later on, Thomson argued that the assumed judgment in the Bystander case—that it is permissible to divert the trolley—is wrong and that the trolley problem, thus, does not exist. Her argument rests on the introduction of a third option to Bystander. According to this option, the bystander can also divert the trolley onto the track on which he himself is standing, which would kill him.

Thomson now believes that it would be permissible for the bystander to divert the trolley to this track (killing himself) or not to hit the switch (killing the five men). Her crucial insight, however, is that it is not permissible to divert the trolley to the other track (killing one person). After all, if you can kill yourself by redirecting the trolley but are not morally obliged to do so, you cannot be permitted to redirect the trolley such that it kills another person. From this, she infers it would also be wrong to redirect the trolley to the side track (killing one person) in the standard Bystander case.[18]

2.16 Always Save the Many? Depends, Says *Scanlon*

There is another, more recent version of deontology, a moral theory called contractualism. Its main proponent is the Harvard philosopher Tim Scanlon. The central idea of contractualism is that of a contractual agreement. Living with others according to rules that one could agree on with them is, itself, something intrinsically valuable. Here is how Scanlon formulates this idea as a principle of normative ethics:

> An act is wrong if its performance under the circumstances would be disallowed by any set of principles for the general regulation of behaviour that no one could reasonably reject as a basis for informed, unforced, general agreement.
>
> *(Scanlon 1999, 153)*

Like Kant's categorical imperative, this principle requires us to examine the principles according to which we make decisions. However, we do not then ask whether it is possible for every other agent to also decide according to these principles. Instead, we ask whether they can be justified in a free agreement with other agents. It is central to consider the objections that other agents would raise against our principles and take them seriously as if we had to reach an agreement.

In moral deliberation, we, therefore, put ourselves in the position of each person who might have an interest in our decision and examine which reasons, from their point of view, speak for and against principles that would permit a certain action. If, for a set of principles, we find that there is an individual whose reasons for rejecting those principles outweigh anyone else's reasons for accepting them, then those principles are inappropriate. If this is the case for any set of principles that would permit a particular action, then that action is morally wrong.

According to *Scanlon*, one of the important tasks of normative ethics is to make clear what is actually important or significant about moral demands. To whom or to what should we actually plausibly feel obligated when we feel morally obligated? Utilitarians usually answer that we are committed to the common good, but Scanlon doubts that this abstract quantity is a plausible source of moral motivation. Instead, Scanlon, like Kant, believes that morality places us in a certain relation to others and that this relation has importance for us. Moral reasoning, Scanlon argues, expresses a reciprocal form of recognition. We treat each other as beings who have important claims on each other—claims that principles for action can also be justified in the light of their perspective—hence the title of his landmark book *What We Owe to Each Other* (Scanlon 1999).

According to contractualism, moral deliberation is based on taking the perspective of different individuals in turn and considering their objections. In such a process, however, only individuals and individual advantages and disadvantages are discussed. None of the individuals can, Scanlon insists, invoke an alleged overall loss of a group. This procedure thus leads to moral judgments that clearly deviate from utilitarianism. Scanlon demonstrates this with a much-discussed example, that of the Transmitter Room:

> Suppose that Jones has suffered an accident in the transmitter room of a television station ... [W]e cannot rescue him without turning off the transmitter for fifteen minutes. A World Cup match is in progress, watched by many people ... Jones' hand has been mashed and he is receiving extremely painful electrical shocks.
>
> *(Scanlon 1999, 235)*

Should we save Jones and spare him an hour of extreme pain? Then we would cause a great deal of frustration to a large number of viewers. Or should we instead make Jones suffer like hell for an hour because it is just one person's pain?

For utilitarians, if the number of viewers is large enough, their frustration at missing a quarter of an hour will outweigh Jones' pain. The example can, therefore, easily be spelled out in such a way that it is not only permissible but even necessary to let Jones suffer for the full hour. Scanlon's contractualism, on the other hand, paints a different picture: We consider Jones and the viewers one by one. The 'collective' frustration of the crowd of spectators is not a variable that can be taken into account. We only consider individual objections. And Jones has the weightiest objection. No individual viewer can (credibly) say that missing a quarter of an hour hurts them more. Scanlon's theory does not only mean that we should save Jones. It specifically means that we should save Jones regardless of the number of spectators. To many, this verdict seems plausible. But does this mean that *Scanlon*'s theory must agree with the view that the number of people involved never counts?

Let us consider the Rock case, a case in which a rescuer can save either the life of one individual who is trapped on one rock or the lives of five individuals who are trapped on a second rock but cannot save the lives of all six. Utilitarianism tells us that the rescuer in the Rock case has a duty to save a greater number of individuals. But contractualism has an individualist restriction that stipulates that each of the five individuals on the second rock can appeal only to their own personal claim of wanting to be saved. Since the individual on the first rock can equally well appeal to their personal claim of wanting to be saved, contractualism seems to yield the result that it is permissible for the rescuer to save either the many or the few.

Surprisingly, Scanlon does not accept this conclusion. He argues as follows: If we only had to choose between two individual persons, it would indeed be correct if we considered both alternatives permissible. But that is precisely why the five people can complain. They can object to the fact that their case is treated as if it were only about one person on their side. So four of them are treated as if they were not there, as if they did not matter at all. Therefore, there is indeed a legitimate objection that the individuals in the larger group can raise: True, no one's individual life carries more weight than another's. But for each of them, the possibility of saving him or her should carry positive weight. And this only happens if the number acts as a tiebreaker.

Not everyone finds this solution convincing. For example, it has been argued that the complaints of the additional agents on the second rock can only break the tie in favor of a duty to save the greater number when considered together with another agent's claim, thus violating the individualist restriction (Otsuka 2006). The debate over how contractualism should deal with the numbers problem continues (for instance, see Parfit 2003; Voorhoeve 2014; Fischer 2021; Henning 2024).

2.17 Conclusion

The First Law of Robotics, introduced by science fiction author Isaac Asimov some 80 years ago in 1942, forbids a robot to injure a human being. The First Law of the ethics of self-driving vehicles is to reduce harm. The German Act on Autonomous Driving 2021, informed by the Ethics Commission's recommendations, includes a provision requiring car manufacturers to develop a collision avoidance tool that effectively avoids, or at least reduces, harm and that prioritizes humans (over animals and things). In cases of unavoidable harm to human life, the Act on Autonomous Driving does not allow certain personal characteristics to be prioritized. However, we learned from the large-scale Moral Machine Experiment that actually a lot of people around the world seem to have a strong preference to save younger over older lives, a somewhat weaker but still marked tendency to spare people with a higher social status (e.g., a doctor in contrast to a criminal, or a business executive vs. a homeless person), and women before men. And indeed, there are historical examples, like triage, or the (unwritten) rule to save women and children first in marine accidents, which seem to make such a strict stance on not differentiating between people at least questionable. Yet we have to be cautious when drawing on empirical studies. While in the Moral Machine Experiment, many people were asked to decide between a man or a woman, a child or an elderly person, etc., other psychologists have found that people might feel forced to answer the questions as they are asked, although they would actually prefer to treat all humans equally.

A puzzling question remains, however. What should be done in dilemma situations in which we have to decide between different numbers of people? Who should die? A single person or rather a group of five? It turns out that legal scholars like the German criminal lawyer Josef Kohler as well as many philosophers, most prominently Philippa Foot and Judith Jarvis Thomson, have thought about this problem for a long time. Despite the caution of many contemporary philosophers, the famous trolley cases can indeed be helpful for thinking about the regulation of moral dilemmas involving self-driving vehicles. Simply put, they are a laboratory of our minds in which we can test what the right thing to do is in reality. They allow us—and you—to inquire about this question. While famous philosophers like Jeremy Bentham and Immanuel Kant have well-grounded arguments for contradictory positions—always saving the many vs. never allowing trade-offs between human lives—Trolley cases can help you to get a better understanding of the moral problem and the pros and cons of the possible solutions. This is important because, as we argue in Chapter 3, what you consider to be the right moral answer should have much more weight in the debate on regulating moral dilemmas involving self-driving vehicles than it currently has.

Notes

1 Section 1e para. 2 No. 2. Note that regulating autonomous vehicle (ethics) is highly dynamic. On 5 August 2022, the European Union enacted the following Commission Implementing Regulation (EU) 2022/1426 (Official Journal of the European Union L 221/1). This regulation lays down rules for the application of another Regulation ((EU) 2019/2144 as regards uniform procedures and technical specifications for the type-approval of the automated driving system (ADS) of fully automated vehicles). In Annex II the following performance requirements are stated:

2. DDT under critical traffic scenarios (emergency operation).

2.1 The ADS shall be able to perform the DDT for all easonably foreseeable critical traffic scenarios in the ODD.

2.1.1. The ADS shall be able to detect the risk of collision with other road users, or a suddenly appearing obstacle (debris, lost load) and shall be able to automatically perform appropriate emergency operation (braking, evasive steering) to avoid reasonably foreseeable collisions and minimise risks to safety of the vehicle occupants and other road users.

2.1.1.1. In the event of an unavoidable alternative risk to human life, the ADS shall not provide for any weighting on the basis of personal characteristics of humans.

2.1.1.2. The protection of other human life outside the fully automated vehicle shall not be subordinated to the protection of human life inside the fully automated vehicle.

2.1.2. The vulnerability of road users involved should be taken into account by the avoidance/mitigation strategy.

2.1.3. After the evasive manoeuvre the vehicle shall aim to resume a stable motion as soon as technically possible.

2.1.4. The signal to activate the hazard warning lights shall be generated automatically in accordance with traffic rules. If the fully automated vehicle automatically drives off again, the signal to deactivate the hazard warning lights shall be generated automatically.

2.1.5. In the event of a traffic accident involving the fully automated vehicle, the ADS shall aim to stop the fully automated vehicle and aim to perform a Minimal risk Manoeuvre to reach the Minimal risk Condition. ADS resuming normal operation shall not be possible until the safe operational state of the fully automated vehicles has been confirmed by self-checks of the ADS or/and the on-board operator (if applicable) or the remote intervention operator (if applicable).

2 Kohler (1915, 432).

3 The specific symbols used might also affect the worldwide comparability of the results given the fact that the written explanations could be displayed when clicked, but this feature unfolded only when participants asked for an explanation.

4 Iserson and Moskop (2007), quoting D. J. Larrey, *Memoirs of Military Surgery, and Campaigns of the French Armies*, Vol. 2 (translated by R.W. Hall; Joseph Cushing publisher, Baltimore, MD 1814), p. 123.

5 Taurek (1977) is a seminal paper on this question.

6 Foot (2003, 29): 'My conclusion is that the distinction between direct and oblique intention plays only a quite subsidiary role in determining what we say in these cases, while the distinction between avoiding injury and bringing aid is very important indeed.'

7 This scenario was presented in an article appearing some ten years after Thomson's first response to *Foot*: Thomson (1985).

8 You might think that this says a lot about the invisible boundaries between academic disciplines. Few philosophers read publications in criminal law, and *vice versa*. However, Thomson did contribute to debates in legal philosophy, and her Bystander case first appeared in an article she

published in the *Yale Law Journal* (one of the most important academic channels for legal scholars). That *Foot* had never heard of Welzel's case, which predates hers by 16 years, might say more about the unwillingness of German criminal lawyers to contribute to the anglophone philosophical debate, and the fact that very few anglophone philosophers can read German (and even if they can, it would be quite a miracle if they were to stumble upon Welzel's article by chance). But to be fair to the great philosopher, it is much easier to find related research these days.

9 Note that the introduction of the bystander (instead of the tram driver used by *Foot*) was intended to show that the bystander would not be responsible for killing the five persons (because they are not the driver of the tram).

10 There she argues that the assumed judgment in the Bystander case—that it is permissible to divert the trolley—is wrong and that the trolley problem, thus, does not exist. The argument rests on the introduction of a third option to the Bystander case. According to this option, the bystander can also divert the trolley onto the track where he is standing (killing himself). Thomson believes that it would be permissible for him to divert the trolley to this track (killing himself) or not to hit the switch (killing the five men) but not to divert the trolley to the other track (killing one person). From this she infers that the latter would also be wrong in the standard Bystander case.

11 Santiago Truccone-Borgogno first suggested a similar scenario to one of us in a philosophical discussion.

12 In her traditional formulation of the trolley problem, Thomson refers to the Surgeon case, which she regards as being similar to the Footbridge case. In that case, a surgeon has five patients with a rare blood type who need one organ each to survive. One needs a heart, one needs kidneys, and so on. By chance, there is a healthy person with exactly that rare blood type. The surgeon could kill the healthy person and take their organs and transplant them into the five patients, saving them. This is just one of the many cases from the medical realm that Thomson and Foot use to illustrate the structure underlying the trolley problem.

13 Some philosophers, however, are not convinced. See, for example, Jenkins, Cerny, and Hribek (2022); Behrends and Basl (2022).

14 In their Supplementary table 2, the numbers concerning fewer vs. many were inadvertently reversed, as Bigman confirmed to us in a personal communication.

15 Against analogies between trolleys and self-driving cars, see Nyholm and Smids (2016).

16 The philosophical discussion as to how *Kant* should be interpreted here, is, however, complex. See, e.g., Varden (2010).

17 This is the position of the German Ethics Commission, for example.

18 For criticism of this argument of Thomson's, see FitzPatrick (2009).

References

Awad, Edmond, Sohan Dsouza, Richard Kim, Jonathan Schulz, Joseph Henrich, Azim Shariff, Jean-François Bonnefon, and Iyad Rahwan. 2018. "The Moral Machine Experiment." *Nature* 563 (7729): 59. https://doi.org/10.1038/s41586-018-0637-6.

Behrends, Jeff, and John Basl. 2022. "Trolleys and Autonomous Vehicles: New Foundations for the Ethics of Machine Learning." In *Autonomous Vehicle Ethics: The Trolley Problem and Beyond*, edited by Ryan Jenkins, David Cerny, and Tomas Hribek, 58–79. New York: Oxford University Press. https://doi.org/10.1093/oso/9780197639191.003.0004.

Bentham, Jeremy. 1977. "A Comment on the Commentaries and a Fragment on Government." In *The Collected Works of Jeremy Bentham*, edited by James Henderson Burns and Herbert Lionel Adolphus Hart. London: Athlone Press.

Berlin, Isaiah. 1969. *Four Essays on Liberty*. Oxford: Oxford University Press.

Bigman, Yochanan E., and Kurt Gray. 2020. "Life and Death Decisions of Autonomous Vehicles." *Nature* 579 (7797): E1–2. https://doi.org/10.1038/s41586-020-1987-4.

Bonnefon, Jean-François. 2021. *The Car That Knew Too Much: Can a Machine Be Moral?* Cambridge, MA: MIT Press.

Bourget, David, and David J. Chalmers. 2020. "Philosophers on Philosophy: The PhilPapers 2020 Survey." https://survey2020.philpeople.org/.

Brown, James R. 1991. *The Laboratory of the Mind: Thought Experiments in the Natural Sciences.* New York: Routledge.

Bruers, Stijn, and Johan Braeckman. 2014. "A Review and Systematization of the Trolley Problem." *Philosophia* 42 (2): 251–69. https://doi.org/10.1007/s11406-013-9507-5.

Černý, David. 2022. "Autonomous Vehicles, the Badness of Death, and Discrimination." In *Autonomous Vehicle Ethics: The Trolley Problem and Beyond*, edited by Ryan Jenkins, David Cerny, and Tomas Hribek, 20–40. New York: Oxford University Press. https://doi.org/10.1093/oso/9780197639191.003.0002.

Cushman, Fiery. 2013. "Action, Outcome, and Value: A Dual-System Framework for Morality." *Personality and Social Psychology Review* 17 (3): 273–92. https://doi.org/10.1177/1088868313495594.

Edmonds, David. 2013. *Would You Kill the Fat Man? The Trolley Problem and What Your Answer Tells Us about Right and Wrong.* Princeton, NJ: Princeton University Press.

Ethics Commission. 2017. "Report on Automated and Connected Driving." June 1, 2017. www.bmvi.de/SharedDocs/EN/publications/report-ethics-commission.pdf?__blob=publicationFile.

Evans, Nicholas G., and Heidi Furey. 2022. "Ethics and Risk Distribution for Autonomous Vehicles." In Autonomous Vehicle Ethics: The Trolley Problem and Beyond, edited by Ryan Jenkins, David Cerny, and Tomas Hribek, 7–24. New York: Oxford University Press. https://doi.org/10.1093/oso/9780197639191.003.0001.

Fischer, Jessica J. T. 2021. "Counting People and Making People Count." *Philosophy* 96 (2): 229–52. https://doi.org/10.1017/S003181912000042X.

FitzPatrick, William J. 2009. "Thomson's Turnabout on the Trolley." *Analysis* 69 (4): 636–43. https://doi.org/10.1093/analys/anp091.

Foot, Philippa. 2003. "The Problem of Abortion and the Doctrine of Double Effect." In *Virtues and Vices*, 19–32. Oxford; New York: Oxford University Press.

Guyer, Paul. 2014. *Kant.* 2nd ed. London: Routledge.

Hart, Herbert L. A. 1982. *Essays on Bentham: Jurisprudence and Political Philosophy.* Oxford; New York: Oxford University Press.

Hauser, Marc, Fiery Cushman, Liane Young, R. Kang-Xing Jin, and John Mikhail. 2007. "A Dissociation Between Moral Judgments and Justifications." *Mind & Language* 22 (1): 1–21. https://doi.org/10.1111/j.1468-0017.2006.00297.x.

Henning, Tim. 2024. "Numbers Without Aggregation." *Noûs* 58 (3): 755–77. https://doi.org/10.1111/nous.12475.

Hilgendorf, Eric. 2021. "Straßenverkehrsrecht der Zukunft." *JuristenZeitung* 76 (9): 444–54. https://doi.org/10.1628/jz-2021-0145.

Howland, Southworth A. 1840. *Steamboat Disasters and Railroad Accidents in the United States.* Worcester: W. Lazell.

Iserson, Kenneth V., and John C. Moskop. 2007. "Triage in Medicine, Part I: Concept, History, and Types." *Annals of Emergency Medicine* 49 (3): 275–81. https://doi.org/10.1016/j.annemergmed.2006.05.019.

Jenkins, Ryan, David Cerny, and Tomas Hribek. 2022. "Introduction." In *Autonomous Vehicle Ethics: The Trolley Problem and Beyond*, edited by Ryan Jenkins, David Cerny, and Tomas Hribek, XIX–XXII. New York: Oxford University Press. https://doi.org/10.1093/oso/9780197639191.002.0006.

Kamm, Frances M. 2016. *The Trolley Problem Mysteries.* Edited by Eric Rakowski. Oxford; New York: Oxford University Press.

Kant, Immanuel. 1996. "Groundwork of The Metaphysics of Morals (1785)." In *Practical Philosophy*, edited and translated by Mary J. Gregor, 37–108. The Cambridge Edition of the Works of Immanuel Kant. Cambridge: Cambridge University Press. https://doi.org/10.1017/CBO9780511813306.007.

Kerstein, Samuel J. 2013. *How to Treat Persons.* Oxford: Oxford University Press.

Kleingeld, Pauline. 2020. "A Kantian Solution to the Trolley Problem." In *Oxford Studies in Normative Ethics*, edited by Mark Timmons, 204–28. Oxford: Oxford University Press. 10.1093/oso/9780198867944.003.0010.

Kohler, Josef. 1915. "Das Notrecht." *Archiv für Rechts- und Wirtschaftsphilosophie* 8 (4): 411–49.

Lübbe, Weyma. 2022. "Corona Triage." *Verfassungsblog*, July. https://verfassungsblog.de/corona-triage-2/.

Mill, John Stuart. 1969. "Utilitarianism." In *The Collected Works of John Stuart Mill*, edited by J. M. Robson, 10: 203–59. Toronto: University of Toronto Press. https://oll-resources.s3.us-east-2.amazonaws.com/oll3/store/titles/241/0223.10.pdf.

Nyholm, Sven, and Jilles Smids. 2016. "The Ethics of Accident-Algorithms for Self-Driving Cars: An Applied Trolley Problem?" *Ethical Theory and Moral Practice* 19 (5): 1275–89. https://doi.org/10.1007/s10677-016-9745-2.

Otsuka, Michael. 2006. "Saving Lives, Moral Theory, and the Claims of Individuals." *Philosophy and Public Affairs* 34 (2): 109–35. https://doi.org/10.1111/j.1088-4963.2006.00058.x.

Parfit, Derek. 2003. "Justifiability to Each Person." *Ratio* 16 (4): 368–90. https://doi.org/10.1046/j.1467-9329.2003.00229.x.

Reeves, Richard. 2008. *John Stuart Mill: Victorian Firebrand*. London: Atlantic.

Robertson-Steel, Iain. 2006. "Evolution of Triage Systems." *Emergency Medicine Journal* 23 (2): 154–55. https://doi.org/10.1136/emj.2005.030270.

Scanlon, T. M. 1999. *What We Owe to Each Other*. Cambridge, MA: Harvard University Press.

Schmidt, Elke Elisabeth. 2022. "Kant on Trolleys and Autonomous Driving." In *Kant and Artificial Intelligence*, edited by Hyeongjoo Kim and Dieter Schönecker, 189–222. Boston: De Gruyter. https://doi.org/10.1515/9783110706611-007.

Taurek, John M. 1977. "Should the Numbers Count?" *Philosophy & Public Affairs* 6 (4): 293–316.

Thomson, Judith Jarvis. 1976. "Killing, Letting Die, and the Trolley Problem." *The Monist* 59 (2): 204–17.

———. 1985. "The Trolley Problem." *The Yale Law Journal* 94 (6): 1395–415. https://doi.org/10.2307/796133.

———. 2008. "Turning the Trolley." *Philosophy & Public Affairs* 36 (4): 359–74.

Varden, Helga. 2010. "Kant and Lying to the Murderer at the Door . . . One More Time: Kant's Legal Philosophy and Lies to Murderers and Nazis." *Journal of Social Philosophy* 41 (4): 403–21. https://doi.org/10.1111/j.1467-9833.2010.01507.x.

Voorhoeve, Alex. 2014. "How Should We Aggregate Competing Claims?" *Ethics* 125 (1): 64–87. https://doi.org/10.1086/677022.

Welzel, Hans. 1951. "Zum Notstandsproblem." *Zeitschrift für die gesamte Strafrechtswissenschaft* 63: 47–56. https://doi.org/10.1515/zstw.1951.63.1.47.

3

WHY YOUR OPINION IS RELEVANT FOR THE LAW AND FOR ETHICS

3.1 Introduction

So far, we have seen that various kinds of moral dilemmas can arise with self-driving vehicles. From the discussion between Professor Principia and Professor Skeptico, we learned that it is very difficult to solve them. It seems that neither of them is absolutely right, and often both provide very good arguments for their way to solve a dilemma. This is a hard nut to crack because there is no way *not* to solve them. We are of the opinion that deciding such hard dilemmas should not exclusively be the job of experts. If it were, morality could only be found in the 'ivory tower' of ethical theory building, which would be strange. Rather, morality seems to have a connection to what (the majority of) laypeople consider the right thing to do, or so we think. What you think about moral dilemmas is more important than you might have guessed before reading this. It is also more important than what most lawmakers and professional ethicists think it is. In this chapter, thus, we wish to take you on another journey. Together we will explore how your opinion and your personal moral preferences can play a role in solving moral dilemmas with self-driving cars. This is important but not exactly easy. If what you consider the right thing to do shall play a role in the discussion, it is also important what your neighbor thinks is the right thing to do. In order to find out what most laypeople think is the right thing to do, psychologists and philosophers run empirical studies. For this journey, the study design for finding morally relevant preferences will become crucial. There are, again, many pros and cons and so we will aim at discovering together the most suitable way to let your opinion and that of your neighbors enter the debate in this chapter.

As we concluded in Chapter 1, in the end, it is our task, as a society, to decide how to program self-driving vehicles for dilemma situations. We have to decide who we prioritize to save and who we put at risk of dying. This is no easy decision and, thus, a great challenge for us as a society. Yet, not making a decision is not an option either because technology gives us the opportunity to program self-driving vehicles in advance for such situations. Hence, deciding not to decide is a decision too.

DOI: 10.4324/9781003485216-3

In Chapter 3, we now not only discuss the relevance of what laypeople like you and your neighbor think about moral dilemmas involving self-driving cars but also show how the moral preferences of laypeople can be aligned and balanced with the view of experts—engineers, ethicists, and lawyers. The result can be seen as something like a recommendation to those who will eventually make that decision, namely democratically elected lawmakers. Let us set off on our journey.

3.2 The Moral Dimension of the Law

We already introduced you to the Moral Machine Experiment in Chapter 2. Studies like this one, even though we had some critical remarks to make, are the most common way to find out what laypeople think about moral dilemmas and how to solve them. There is, however, strong opposition to using such studies for our purposes. Professor Principia is not a big fan of attempts to study what laypeople think about moral dilemmas. In her pronounced criticism of such attempts, she rejects any moral value of experiments such as the Moral Machine.[1] She considers the main problem of the experiment to be a large-scale trivialization of the relevant issues. To her, its authors ignore the relevant moral and legal aspects on several occasions. Therefore, she thinks that the Moral Machine, the website where everyone can decide on moral dilemmas with self-driving cars, and other similar experiments do not contribute to the ethical debate and the foreseeable legal regulation of self-driving cars.

An important point that Professor Principia makes is that she wonders whether the researchers behind the Moral Machine Experiment were somehow careless in their handling of the law. There are two famous cases in the English legal system which both deal with moral decisions about life and death. Hence, life-and-death issues of 'innocent road users' should not be debated isolated from the law.

In the first historical landmark case (R v. Dudley and Stephens [1884] 14 QBD 273 DC), two shipwrecked sailors had killed and eaten their cabin boy, who had fallen into a coma. Despite acting out of necessity to save their own lives, both were charged with murder. The researchers behind the Moral Machine Experiment take the authorization for granted that a self-driving car could kill innocent people in the case of necessity, Principia says. To her, it is not an option to even consider that innocent road users may be injured or even killed when the alternative involves any risk to the passengers of self-driving cars. When even the poor shipwrecked sailors were charged, how can one even start to think about programming for self-driving cars that deliberately kill innocent people to save other innocent people just because most laypeople prefer such programming?

Professor Principia highlights yet another important case, namely that of the separation of the Manchester Conjoined Twins in 2000. Medical evidence showed that one of the conjoined twins had a primitive brain, very little heart function, and no functioning lungs, surviving only due to a shared common artery supplied by the other twin. If separated, the stronger twin was assessed as having a very high survival rate, but the weaker one was almost guaranteed to die. If left conjoined, their life expectancy was estimated to be around six months. Reporting this case, Principia emphasizes the complexity of the legal process which is necessary to deal with life-and-death decisions. Only after a long and dramatic trial did the Court of Appeal for England and Wales rule that it would be permissible to sever and thus kill the weaker twin in order to save the stronger one (Re A [2001] 2 WLR 480). Regardless of what one thinks of this judgment, her point is that such dilemmatic situations cannot be solved by public referendum but are in need of a careful legal analysis carried out by independent experts such as judges.

In light of these two dramatic cases, you might think that Professor Principia's position is appealing. Our leading idea that what the people say should play a role was perhaps promising but is not compelling after all. You, dear reader, should probably not have much say in who survives accidents with self-driving cars. The experts will carefully deliberate and let you know your chances. However, Professor Skeptico is not that easily convinced. Principia's position needs to be treated with caution.[2]

In the case of the shipwrecked sailors, Principia makes the point that one should not regard the moral preferences of laypeople as relevant for moral theorizing because they might violate the law: Laypeople might prefer programming for self-driving cars that risk the lives of innocent people, although even the shipwrecked sailors were charged with murder. This point is problematic for several reasons, says Professor Skeptico. He has several arguments which speak against this criticism.

First of all, the law is not static but open to changes orchestrated by a (democratic) law-making process. In principle, any preference supported by a majority might become law. Yet, Professor Principia interrupts, in constitutional democracies, new laws must not violate constitutionally guaranteed fundamental rights and the right to life certainly belongs to this guarantee. And in Common Law systems—such as the English legal system—courts are bound by precedents. This is a very effective way to establish legal doctrines step by step, case by case. Professor Skeptico is not very convinced by this intervention. The problem is that it is notoriously difficult to determine whether a particular precedent applies to the case at hand or not. For instance, in the case of the Manchester Conjoined Twins, the courts disagreed on precisely this question.

And even if a precedent applies to a new case, there is a related problem: 'If the only tool you have is a hammer, everything looks like a nail.' Arguing from precedents notoriously risks drawing analogies between cases that are, in fact, not analogous because the precedent is the only tool you have, explains Skeptico. Both cases referred to by Principia are emergency situations. Not acting would have meant the shipwrecked sailors starving to death or letting both twins die. The regulation of self-driving cars is quite different. The decisive difference, says Skeptico, is that it does not concern individual decisions in emergencies but rather the general regulation of situations that will very likely occur once the technology for self-driving cars is sufficiently advanced. It is, thus, important to discuss the regulation of moral dilemmas in traffic accidents involving self-driving cars.

In other words, the technological innovation of self-driving cars and their computational capacity puts us in a new situation. Computational capacity far outstrips human capacity when reacting in an accident, and this offers new potential for regulation. Due to the cognitive limitations of humans, in traffic accidents, they mostly react fast and affectively instead of thinking through all possible options. While we could claim that the moral dilemmas of self-driving cars should be solved by existing legal provisions, we might also acknowledge that this genuinely new situation demands new regulation.

A second point worth noting, adds Skeptico, is that moral preferences must not necessarily match the law. While it might be moral to comply with the law, it is not automatically amoral or immoral to have preferences that violate the law. Just as there are laws that are amoral (e.g., traffic laws stipulating on which side of the road cars must drive), there are certainly also immoral laws, i.e., laws that are immoral but still count as law. For example, think of some contemporary migration laws or tax laws, for instance, or a particularly severe case, the 'Nazi laws,' which were so immoral but were nevertheless regulations formally adopted in law-making procedures and courts (albeit in a fascist regime, which led many legal philosophers to hold that they do not even deserve the label 'law' and should not be regarded as binding). Thus, we need reasons why some preferences are immoral. Simply stipulating that they would be immoral because of a conflict with the law is not convincing. Without providing further reasons, it is circular to say that some preferences are immoral because they might violate the law.

Or think about it this way, Skeptico says provocatively: Is it not more plausible to hold that a law that is in conflict with moral preferences should be changed? Our conceptual understanding as to what morality is might well be informed by what the law prescribes but it should not be determined by the law.[3]

3.3 The Legal Notion of Necessity

Professor Principia does not want to give up her law-based argument that fast. She reiterates her point and refers to criminal law: It was even illegal to kill the innocent cabin boy in order to save the lives of the other shipwrecked sailors. How can this be legal in relation to self-driving cars then? At least in English law, she says, there is no defense of necessity to charges of murder. And choosing to kill innocent road users (even when other road users are then saved) is nothing but murder!

This is a heavy charge. Nevertheless, using the example of English (criminal) law as a general argument against the necessity case is problematic, replies Professor Skeptico. First, the general assessment is not entirely correct anymore. It was the very Manchester Conjoined Twins case—to which Professor Principia referred to above herself—in which the courts departed somewhat from the long-standing precedent that there is no necessity defense against murder charges (Santoni de Sio 2017, 416). This development shows the importance of the point Skeptico highlighted before: The law is not static but open to change. Most laws are amoral or moral but others are immoral—and the ratio between the latter two changes constantly. What is moral and what is immoral cannot be determined by asking what the law says.

Second, a slightly different image of the circumstances and potential arguments arises when we turn to Continental criminal law doctrine. Consider Section 34 (necessity as justification) and Section 35 (necessity as defense) of the German Criminal Code. In emergency situations (in the case that two lives are in danger but neither of the endangered persons engages in an illegal attack endangering the life of the other person), the sacrifice of human life is not justifiable. So far so good. However, Professor Skeptico has also read progressive German criminal law scholars and points out to Professor Principia that an act resulting in saving person A and killing person B might be justifiable (or excusable) out of necessity (Hilgendorf 2018, 60–70).

There is a famous example to demonstrate this, already mentioned in Chapter 1, namely the 'Carneades case,' which is similar to that of the shipwrecked sailors, and the classic example in Continental criminal law doctrine. Professor Skeptico reminds us that this scenario goes back to the Carneades of Cyrene (second-century BC). This time, two shipwrecked sailors see a plank that can only support one of them. One sailor reaches the plank first; by the time the other reaches the plank, he is about to drown. Is the surviving sailor to blame if he pushes the first sailor off the plank, causing him to drown?

Following legal doctrine, the action of the shipwrecked sailors killing the cabin boy in order to save themselves has to be excusable because if their behavior was forbidden, the law would require the sailors to sacrifice themselves. As this is too much to ask of individuals, in Germany such behavior is, in legal terms, considered 'excusable,' i.e., it remains illegal but goes unpunished (for a general overview, see Dreier 2007).

The final blow, however, for arguments from criminal law doctrine against specific programming of self-driving cars in the case of moral dilemmas is that criminal law is the wrong legal discipline to consult. Criminal law is—in a nutshell—about individual culpability in relation to a specific deed that is considered wrong. Thus, a person who commits murder is to be punished. This is fundamentally different, says Professor Skeptico, from the challenge we face with self-driving cars. Our task is not to decide whether after a particular accident someone's action was relevant under criminal law. It is not about someone causing an accident because they were drunk. In our case, we want to find the best programming for situations in which we neither know the occupants of the self-driving car nor other road users. Our goal is to make the streets as safe as possible and to reduce harm as much as possible. The programming that is considered best shall be adopted by the lawmakers and guide car manufacturers as much as individuals before the first accident happens. Hence, we do not look at a past action needing to find the culprit. We are looking into the future and want to find programming which is morally permissible.

This makes Professor Principia think. Perhaps criminal law is not her best line of argument indeed. However, Professor Skeptico just said that he wants to find morally permissible programming. This is going to be hard for him because it seems to be questionable to Professor Principia what concept of morality Professor Skeptico has in mind. To her, it seems that he lacks any sound concept of morality.

3.4 The Concept of Morality

Professor Principia leaves law behind her. This was anyhow more of a vague point of interest rather than her profound area of expertise. The discussion was nevertheless useful, she thinks, because this discussion brought to light that the Moral Machine Experiment, which Professor Skeptico is generally in favor of, lacks any sound concept of morality. The problem, Principia notes, is less the attempt by the researchers to take public morality as the foundation for legal and ethical frameworks but more the reductionist image they draw of public morality, ignoring its embeddedness within a historical and cultural process. Public morality, Principia says, is not a spontaneous beast. Rather it has evolved over a lengthy period of time. History, art, literature, or indeed any form of culture play an important role, which seems not to have been taken into account by the approach of this experiment. The researchers behind the Moral Machine Experiment simply equate the preferences of their participants with expressions of public morality, or so she claims. However, nothing is more repugnant than the majority, says Professor Principia (loosely quoting Goethe). The fact that the majority holds a specific position does not make the earth flat (in case flat earthers become the majority again) nor does it make immoral positions moral. Casting votes is, thus, rather an inappropriate tool to determine public morality.

Following Principia, studies such as the Moral Machine Experiment intend to replace a considered discourse on complex and controversial ethical problems with simple yes–no questions answered by uninformed individuals. For her, this methodology is highly unreliable: 'Public morality' in this context is nothing but ill-informed, unconsidered preferences provided by participants sitting in front of their computers, clicking 'yes' or 'no,' mostly instantly and likely thoughtlessly, much like playing a computer game. Therefore, she says, the experiment is not only amoral—it is even immoral. It promotes a simplified and wrong impression of public morality, eliminating the distinction between moral judgments and personal prejudices.

Jean-François Bonnefon, one of the architects of the Moral Machine Experiment, disagrees, as Professor Skeptico happens to know, having read Bonnefon's book. In the book, Skeptico explains, the rationale and context for the experiment is given by the researchers themselves. Their idea behind the Moral Machine Experiment was not to decide what ought to be done, all things considered, by polling the general public on their moral preferences. With reference to the German Ethics Commission, one of the researchers writes: 'there is nothing wrong with trusting well-informed specialists, but it is unfortunate that citizens were not given an opportunity to voice their preferences, especially when the specialists disagreed' (Bonnefon 2021, 71).

It is, of course, a grand question to ask what morality is, and it is unlikely that our two philosopher friends will find much common ground on this issue. The criticism that morality cannot be defined using an experiment like the Moral Machine hinges as much upon answering this question as upon the experiment itself. The big issue is the question of where we are to find morality. Is morality only to be found in the 'ivory tower' of ethical theory building or is it (also) connected to what (the majority of) laypeople consider the right thing to do? We see that an important point in our discussion is also a controversial point among philosophers. If the latter has to play a role, if your opinion, dear reader, ought to be included in the programming, then the study design for finding morally relevant preferences becomes crucial. While it is correct to demand high methodological standards when investigating preferences, especially moral preferences, it is all too easy to write off any and every experiment aiming to shed light on public morality. Because this is so important, we will discuss this problem in more detail in the following sections. Again, the details are controversial, but in this case, the details are very important to get closer to an adequate answer.

Claiming that majorities do not necessarily make morally right judgments does not mean that they are most likely to arrive at wrong moral judgments. Certainly, majorities are not a guarantee for morality—but neither are the judgments of individuals, even those of moral philosophers. Professor Principia seems to endorse a concept of morality that requires a person to possess a certain level of expertise, or at least to be sufficiently well informed, in order to make moral judgments. She criticizes the Moral Machine experiment for considering ill-informed, unconsidered preferences. It is Principia's disregard for the opinions of laypeople in contrast to what experts think is right—even when they disagree among themselves—that Bonnefon addresses. Would it not be better to consider what the people say instead of relying on experts only?

Recent empirical research on the intuitions of professional moral philosophers has challenged the idea that there is such a thing as moral expertise in the sense that the moral intuitions of moral philosophers are more reliable than those of laypeople (for an overview, see Horvath and Koch 2020; Paulo 2020b). As the philosopher Eric Schwitzgebel and the psychologist Fiery Cushman put it in their study of philosophical expertise, 'if there is a level of philosophical expertise that reduces the influence of factors such as order and frame upon one's moral judgments, we have yet to find empirical evidence of it' (2015, 136). This is not to say that philosophers are useless and have no expertise whatsoever. The point is merely that they seem to be prone to some of the same biases that affect the reasoning of all human beings, even within their field of expertise. And such biases surely should not be included in ethical decision-making. Other studies suggest that philosophers also tend to underestimate how easily their judgment can be distorted by irrelevant factors such as the mere framing of problems or the order in which certain scenarios are presented.

Summing up, arguments for a total rejection of the Moral Machine Experiment (or similar studies) for normative reasons—as defended by Professor Principia—are not convincing. In response to her critique, Professor Skeptico argued that neither from a legal nor from an ethical perspective is there a point of concern that convincingly undermines the *raison d'être* of the Moral Machine Experiment. Still, it does have to be stressed that investigating public morality is a sensitive issue that requires a critical evaluation of the methodological approaches throughout the process. We need sensible ways of accounting for both public morality and traditional moral theory.

3.5 The Practical Relevance of 'What the People Think'

It is certainly not enough to claim the irrelevance of 'what the people think' for what is morally right to make that irrelevant, says Professor Skeptico in a self-confident manner. But in what way exactly is it relevant for moral questions, then? This is a very good question to ask indeed. Discussing the potential relevance of their findings for public policy, and in the context of potential trade-offs between the young and the elderly, the researchers behind the Moral Machine Experiment write:

> This does not mean that policymakers should necessarily go with public opinion and allow autonomous vehicles to preferentially spare children [...]. But given the strong preference for sparing children, policymakers must be aware of a dual challenge if they decide not to give a special status to children: the challenge of explaining the rationale for such a decision, and the challenge of handling the strong backlash that will inevitably occur the day an autonomous vehicle sacrifices children in a dilemma situation.
>
> *(Awad et al. 2018, 60)*

And they are not the only ones who believe that such findings are relevant. Indeed, some thinkers consider them to be of prime importance for questions of justice. Professor Skeptico points to some prominent philosophers. For instance, David Miller holds that any theory of justice is 'to be tested, in part, by its correspondence with evidence concerning everyday beliefs about justice' (1999, 51). But, Professor Principia asks, what is the point of taking evidence of laypeople's beliefs into account? There are practical and theoretical reasons for doing so, replies Professor Skeptico.

Let us begin with the practical reasons for philosophers to regard empirical evidence of 'what the people think' as relevant. These reasons concern the very essence of the work of moral philosophers. As Miller puts it:

> The final aim of normative theorising as carried out by philosophers is to provide practical guidance on matters of moral or political concern. When such a theory is presented, the principles it contains must, therefore, be ones that people might actually come to embrace and act upon. But if so, they must be accessible to the relevant agents, which means that there must at least be a bridge between these principles and the beliefs (about justice and so forth) that people already hold. [...] So, it is important to find out, at least as a starting point, what these beliefs actually are.
>
> *(2020, 274)*

This might sound obvious. But the issue is contested as to what the relation between academic normative theorizing and practical applications is. Miller's colleague Jonathan Wolff argues against those who think that abstract theorizing should take center stage:

> Of course [...], radical philosophical arguments are a vital part of the debate, and add to the stock of ideas that enrich discussion. But on their own they will settle nothing. The methodology [...] recommended here suggests that when thinking about a practical issue, we should start at the other end: not at the philosophical theories but current disagreements in the public policy area. We need to ask: what do people think they disagree about? And is that the best way of understanding their disagreement? Is there a better way? And if so, does that open up new avenues for making progress?
>
> *(2019, 35–36)*

The point is, explains Professor Skeptico, that the philosopher is 'essentially a democratic underlabourer' (Swift and White 2008, 54). They can offer their normative arguments in public discourse—but when they do this, their role is essentially the same as that of other citizens. Professor Skeptico goes on: The claim of philosophers to have special expertise seems to be limited to providing a framework for the moral discourse. This framework exercise consists of offering help in formulating principles and policies, indicating their implications, offering criticism, more arguments, and justifications, etc. so that public discourse about a public policy can flourish. Their role, however, is not to dominate such discourse.

These remarks already indicate, Professor Skeptico continues, that even for those philosophers who think that they should pay attention to what the people say, public opinion does not determine which policies should be adopted. However, Professor Principia intervenes: What about preferences that should definitely not inform public policy? Take, for instance, studies which show that people have a strong preference to save members of their own ethnic group over those of other ethnicities (e.g., Swann et al. 2010; Cikara et al. 2010). Is this public morality? In addition, it is quite likely that similar patterns could be found for other forms of discrimination, says Principia. Just imagine what faithful people's preferences might look like in the face of non-believers. This thought could be extended. Use your imagination! So the worry is that individuals actually have discriminatory preferences instead of morally valuable preferences, and that paying attention to what the people say might lead to discriminatory and, thus, to bad policies. Professor Skeptico gives in and says that indeed, for these reasons, it is important to acknowledge that getting to know what the people think is only the beginning of the work of philosophers interested in issues of public policy. But he urges Principia to bear with him as he tries to explain how this might work. We can actually filter out biased and discriminatory preferences!

3.6 How to Combine What You Think Is the Right Thing to Do With the View of Moral Experts

The reason for taking your and other laypeople's beliefs into account concerns the appropriateness of what is morally right or wrong, all things considered. It is about the core of ethics, namely to find out what is the right thing to do, says Professor Skeptico, full stop.

Skeptico points again to philosopher David Miller, who uses a method called 'reflective equilibrium'[4] to combine laypeople's moral preferences with classic moral theories as advanced by philosophers. This notion was developed by John Rawls, one of the most influential philosophers of the twentieth century, in his landmark 1971 book *A Theory of Justice* (Rawls 2005). In Rawls' account, the method is carried out individually. One starts from one's *intuitions* about certain moral problems or particular cases and aims to systematize those intuitions with the moral principles one holds to be true. In that process of systemization, intuitions and principles are reconsidered and revised until the set of intuitions and principles is coherent. Acknowledging its widespread use in philosophy, Arras has called the method of reflective equilibrium '[t]he way we reason now' (2007); some philosophers even believe it is 'the only defensible method: apparent alternatives to it are illusory' (Scanlon 2002, 149).[5]

The intuitions relevant to this method are not mere hunches or gut feelings. Rather, reflective equilibrium relies on a certain kind of considered judgments, namely those that are 'likely to arise in ordinary life … in order that the judgment in question be made in the effort to settle problems with which men are familiar and whereupon they have had an opportunity to reflect' (Rawls 1951, 182). What is more, these judgments are not derived from general moral principles. In other words, these intuitions are reflected judgments on particular cases; individuals must be very confident that they are correct, for instance, because they had all the relevant information and the chance to reflect on them in various situations and under favorable circumstances (without being distracted, agitated, or angry, say), and they must not be inferred from principles, i.e., their epistemic weight lies in their immediate appeal, not in abstract theorizing.

For Miller, empirical evidence of 'what the people think' is relevant in reflective equilibrium because such evidence can reveal which principles of justice laypeople endorse. This, in turn, can help to assess the extent to which one's own judgments about justice qualify as intuitions in reflective equilibrium. Now imagine, Professor Skeptico invites Professor Principia to ponder, if you were to learn that your judgment about a particular case conflicts with a widely held justice principle, what would you do? Would this not give you a reason to inquire whether your own judgment is not sufficiently reliable, possibly because it results from your particular cultural or social position or some other reason? Just think of the research carried out by the anthropologist Joseph Henrich. He demonstrated that WEIRD people, an acronym which stands for those coming from the *Western* world, being *e*ducated in the *i*ndustrialized world, *r*ich compared to the rest of the world, and living in a *d*emocratic society, consider different actions to be the right thing to do in contrast to people from the rest of the world (Henrich 2020). Consider this example, says Skeptico. For most WEIRD people, it would be morally wrong to lie for a close friend who exceeded the speed limit just to help your friend avoid a legal penalty. For most people in the rest of the world, being loyal to a close friend appears to be more important than an abstract concept of honesty and responsibility. Learning about such a conflict between their judgment and a justice principle held widely by people in the rest of the world might also prompt them to examine whether or not their judgment has resulted from an adjustment to a principle that they (as a WEIRD person) hold (which it should not, of course, if we are looking for a universal principle).

Perhaps inspired by Miller, a group of ethicists has recently suggested a sensible method of accounting for both public morality and traditional moral theory. What they call 'Collective Reflective Equilibrium in Practice (CREP)' might provide common ground between Professor Skeptico and Professor Principia. The authors of CREP explain the reasons behind their model as follows:

> We cannot simply base our policy on the strongest public intuitions; nor can we simply ignore public intuitions in favour of top-down policy-making. The challenge, then, is to find a way of integrating ethical theory and data about the spread of public intuitions while minimizing the risks associated with each.
>
> *(Savulescu, Gyngell, and Kahane 2020, 656)*

The method aims to combine moral theorizing by experts and empirical findings about public preferences from laypeople. There are not only practical reasons to take such preferences into account. They can also provide justification for moral views and public policies based on these views because the fact that a moral view is widely held speaks for that view, as does the fact that it is held by members of different cultures (Huemer 2008). After all, these are useful heuristics to make sure that an individual's moral views are not culturally biased. The example showing the difference between WEIRD people and the rest of the world was just one illustration of cultural differences. Therefore we must be very careful when empirical studies claim universal results. However, if you learn that your opinion on a particular problem conflicts with what people worldwide overwhelmingly believe, this is a reason to reconsider your opinion (if you want to come up with an acceptable moral view).

Standard reflective equilibrium is carried out individually. Every philosopher tries to reach the state of reflective equilibrium between moral intuitions and more abstract principles. The CREP method, in contrast, describes a collective enterprise. CREP does not start from an individual's intuition about certain moral problems or particular cases. Instead, the input comes from data about public moral preferences.

3.6.1 Careful, Some Preferences Might Not Be Reliable!

The first thing Professor Principia notes is that your preferences and those of others may be quite different from what Rawls had in mind for his method. The conditions he set for standard reflective equilibrium are much higher. He requires that intuitions are well-considered and made under favorable conditions. Surely very few people who 'played' with the Moral Machine Experiment took their time and provided thoroughly considered judgments. Professor Skeptico agrees: That is true, but it does not mean that the preferences collected in empirical studies are worthless. When Rawls demanded that the intuition input should be well considered and made under favorable conditions, he had specific reasons in mind. His idea was meant to ensure that the intuitions are as epistemically robust as possible. He wanted to avoid the 'garbage in, garbage out' problem. The output of the method will be useless if the input is unreliable. To ensure that, he was looking for a way to get at intuitions that are truly our sincere intuitions and not merely unreflected shots from the hip. The important point is that there are ways to ensure robustness in the moral preferences collected in empirical studies. Psychologists have developed a very strict methodology in order to ensure that their data are replicable. This means that if you do exactly the same study again, you will receive the same results, and not different preferences for exactly the same situation.[6] For instance, one important way of distinguishing robust moral preferences from unreliable preferences is to filter out those preferences that are not representative. Hence, certain statistical requirements have to be fulfilled. It is, for example, not methodologically correct to carry out a study with too few participants. In addition, if researchers want to claim that the moral preferences found are representative of all citizens, it is not reliable to conduct a study with young participants only but rather with young and old participants. Think of the moral dilemma about whether a self-driving car should save the young or the elderly. The result is quite likely to be different if you conduct a study with only older participants. Another point is that it is important to sort out unreliable preferences that are sensitive to irrelevant and potentially disturbing factors.[7] It was in the 1970s that psychologists Amos Tversky and Daniel Kahneman introduced the term 'cognitive bias' (Tversky and Kahneman 1974), explaining that people use heuristic principles to reduce complexity when making decisions. While generally this is fine, they showed that it can cause people to commit severe and systematic errors. Consider only the first of the examples they present, the representativeness heuristic: 'Steve is very shy and withdrawn, invariably helpful, but with little interest in people, or in the world of reality. A meek and tidy soul, he has a need for order and structure, and a passion for detail.' After this brief description of Steve, they ask you to take one minute, to read it again and then please order the probability that Steve is engaged in a particular occupation from this list: Steve is a (1) farmer, (2) a salesman, (3) an airline pilot, (4) a librarian, or (5) a physician? Now we ask you to pause for a moment and take the time to answer this question.

Are you ready? What order did you put them in? Does the librarian feature at the top of your list? If so, you fell short of the representative heuristic, too, because the similarity of Steve's description with the stereotype of a librarian influences your decision too strongly. For example, explain Tversky and Kahneman, 'the fact that there are many more farmers than librarians in the population should enter into any reasonable estimate of the probability that Steve is a librarian rather than a farmer.' Since the 1970s, a whole range of demographic factors and cognitive biases have been discovered that are irrelevant and potentially disturbing.

Let us continue with demographic factors. Some studies report *gender* differences in response to trolley cases as discussed in Chapter 2 (Petrinovich, O'Neill, and Jorgensen 1993; Zamzow and Nichols 2009). For example, men tend more toward utilitarian intuitions, finding it permissible to cause harm to prevent greater harm (Fumagalli et al. 2010; Bartels and Pizarro 2011; Friesdorf, Conway, and Gawronski 2015). *Personality*, too, seems to play a role: Bartels and Pizarro (2011) found that where people score on the psychopathy scale partly predicts their judgments in a range of moral cases. Gleichgerrcht and Young (2013) found that people who respond to trolley cases in a utilitarian manner have less empathy than those who do not (see also Kahane et al. 2015). Banerjee, Huebner, and Hauser (2010) show that *religiosity* and *political orientation* have effects on moral intuitions in a variety of dilemmas, especially in sacrificial dilemmas (Machery 2017, 67). Some studies did not find variations with respect to *cultural differences* between the test subjects (Hauser et al. 2007; Abarbanell and Hauser 2010). But some experiments did. Ahlenius and Tännsjö (2012), for instance, report that American, Russian, and Chinese test subjects responded systematically differently to variants of the trolley cases. Gold, Colman, and Pulford (2014) report similar effects for British compared to Chinese test subjects.

This is not the end of the list. One has to consider cognitive biases as well. The *mood* of test persons seems to affect their intuitions (Valdesolo and DeSteno 2006). They tend to react differently with an induced positive mood—through watching a comedy clip, for instance—compared to when they are in a neutral mood (see also Strohminger, Lewis, and Meyer 2011). When an intuition varies depending on the order in which two or more cases are presented, we speak of an *order effect*. That is, test subjects have different intuitions about two cases, C1 and C2, when they see C1 before C2 compared to when they see C2 before C1. Several studies (Lanteri, Chelini, and Rizzello 2008; Wiegmann, Okan, and Nagel 2012; Liao et al. 2012) revealed a number of order effects in trolley scenarios. One example is the famous 'Loop case' devised by philosopher Judith Jarvis Thomson. It seems as if intuitions about the Loop case vary according to the order in which the scenarios are presented (Liao et al. 2012). Depending on which scenario was presented before the Loop case, participants agree or disagree that it is morally permissible to divert the trolley onto a second track where there is an innocent bystander in order to save five innocent people on the track where the trolley is traveling. This is an important insight because this Loop case is designed in such a way that the trolley would be stopped by hitting the innocent bystander and would otherwise continue, with this second track looping back toward the five people (hence the loop case). So killing the innocent bystander is necessary to save the five. Petrinovich and O'Neill (1996) found that *framing effects* can be quite significant: Test subjects react differently to trolley cases when they are formulated in terms of saving one person compared to killing five. That is, a significant proportion of test subjects' intuitions vary with the wording of the scenarios. Finally, people react differently to trolley cases when they are presented in their native language compared to a foreign language (one they understand, obviously) (Costa et al. 2014).

This long list is by no means complete, and very likely some biases have not yet been discovered. Considering all this, how can one still say that the moral preferences of laypeople should play a role? Well, think of the CREP method which takes account of this list and proactively advocates for the need to filter out unreliable preferences. When intuitions vary with demographic factors or with the cognitive biases just mentioned, they should not factor in the CREP model because they are unreliable. As the list of potentially disturbing factors shows, many of them have been found to play a role in trolley cases. That is, when people were asked to state their intuitions about the various versions of trolley cases, their responses varied with gender, personality, religiosity and political orientation, culture, order, and framing. This sounds as if we really should not rely on the preferences of laypeople about trolley cases. And since we argued that there are strong similarities between the programming of self-driving cars for unavoidable accidents and trolley cases, it would not be far-fetched to imagine that the unreliability carries over to findings about public preferences concerning self-driving cars. This is a reasonable worry, but things are a bit more complicated than that. One can remove unreliable moral preferences so that only reliable intuitions are fed into the CREP model. We will explain how in a minute.

3.6.2 Sorting Reliable From Unreliable Preferences

So how does one separate reliable from unreliable preferences? The trick is to check whether or not individual studies are well-designed. The wealth of biases and other problems related to methodology pointed out force us to check that we are only consulting studies that live up to the highest scientific standards. Since the demographic factors and cognitive biases listed above have been studied intensively and are well-known among psychologists and philosophers, new studies are expected to be designed in such a way that these influences on people's intuitions are avoided. The study will only be published by a respected journal if it was designed to avoid biases and other irrelevant influences.

Ok, this is common ground of course, says Professor Principia, but is it not the case that the Moral Machine Experiment you referred to before has some important limitations to this end? You just admitted that demographics are important, and if I got it right, it seems that this experiment's data set is skewed toward males in their 20s and 30s, which raises the question as to whether the findings merely reflect the preferences of young men, rather than being universal. Good point, says Skeptico. However, the researchers investigated the role of demographics in the pattern of preferences they observed. They found that 'individual variations have no sizable impact' (Awad et al. 2018, 60). So there is good reason to believe that demographics did not unduly influence the findings.

But what about the framing effect, Principia asks? Well, Skeptico replies, the study relied on both written and pictorial descriptions of the ethical dilemmas. Some used only written descriptions and some used only pictorial ones, yet no difference was found between groups that relied on the one rather than on the other. In order to rule out framing effects, in the written version, the scenarios and answer options were formulated in the same (almost identical) terms.

Fine, but I recall the work of the psychologists Bigman and Gray (2020) who you mentioned earlier (in Chapter 2). If I remember correctly, they investigated whether participants' responses would be different when a third option was added to the mere possibility of killing either A or B. Did they not initially ask participants if the car should be programmed to kill A, to kill B, or to treat the lives of A and B equally? This formulation might lead to a framing effect because the 'killing' wording is very different from the 'treating equally' wording. The latter seems way more appealing to me. I would rather treat people equally than kill them. I assume that their participants felt the same. Yes, that is true, Skeptico answers. But precisely in order to avoid this framing effect, which was actually present in one of their studies, they reformulated the answer options and ran a new study that avoided this framing effect. The new study showed the same main results. So no need to worry about framing here.

To anticipate another critical remark by Principia, Skeptico adds that when asking for responses to different versions of trolley-like cases, it is also imperative to randomize the order in which the different versions of moral dilemmas appear. In this way, order effects cannot influence the outcome. The Moral Machine Experiment and Bigman and Gray's study did, of course, follow this imperative.

For valid studies, it is necessary to avoid the well-known and potentially disturbing demographic factors as well as cognitive biases. Only if they do can they offer reliable findings. Sure, no study is perfect, and very likely we will discover additional factors and biases as already mentioned which influence decisions in a way that should be sorted out. Nevertheless, so far, we have clear criteria for how well-designed a study is regarding potentially disturbing factors. The CREP method uses these criteria to sort reliable from unreliable moral preferences.

The screening process may also include comparisons between stated preferences and revealed preferences. Sometimes, survey data do not properly reflect what people actually do. Before we consider an example, it would be great if you could take a minute and rate your personal driving skills for us. On a scale from 1 to 10, what kind of rating would you give yourself? '1' is very poor, '5' is average, and '10' is excellent. Now let us guess. We assume that you gave yourself a 7, possibly even an 8. While you probably do not think of yourself as being another Lewis Hamilton or Max Verstappen, you are very likely to overestimate your skills, because this is what the majority of car drivers do. Most drivers think that their driving is above average. This is statistically impossible, of course. Research in the United States and Sweden shows that 88% of U.S. drivers and 77% of Swedish drivers 'believed themselves to be safer than the median driver,' and 93% in the United States and 69% in Sweden 'believed themselves to be more skillful drivers than the median driver' (Svenson 1981, 146). This phenomenon is not limited to driving. Overconfidence is found in many areas. We do not want to go off at a tangent, but if you are a student, this might be of interest to you: '80% of students think they will finish in the top half of their class' (Hoffrage 2022).

However, especially in the context of novel technologies like self-driving cars, it is hard—if not impossible—to gather information on revealed preferences. But researchers can try to design their studies in such a way that makes the choice situation as realistic as possible. For instance, some studies have used virtual reality to simulate real-life decision-making in sacrificial dilemmas, finding that people's responses tend to be more utilitarian in virtual reality scenarios compared to abstract descriptions of the same scenario (Bostyn, Sevenhant, and Roets 2018; Francis et al. 2017); this has also been done in the context of self-driving cars (Sütfeld et al. 2017).

Another factor is replication. If a finding is sufficiently interesting, other researchers will try to reproduce it with a different sample of test subjects and often with a slightly altered design that should not affect the outcome. If they can replicate the study's main findings, this lends further support to the findings. If they cannot, this casts doubt on the findings.

So were other researchers able to replicate the findings of the Moral Machine Experiment, Principia asks? Well, Skeptico says, at least the main findings were successfully replicated in later studies (e.g., Bigman and Gray 2020; Bodenschatz, Uhl, and Walkowitz 2021). However, as mentioned before (in Chapter 2), researchers found a design flaw in that large-scale experiment, namely that it forces people to decide between two bad options when they actually might want to treat all potential victims equally. Bigman and Gray (2020) found that with the additional option to treat all potential victims equally, the preference to save children over the elderly dropped from 96.1% to 38.8%, while 61.1% preferred the equality option to treat children and the elderly alike. Similarly, the preference to save those with high social status drops from 79.7% to just 12.7%. The vast majority (84.7%) want to treat potential victims with high and low social status equally. Other studies found similar patterns (e.g., De Freitas and Cikara 2021; Bodenschatz, Uhl, and Walkowitz 2021).

The CREP method suggests 'laundering' moral preferences along the lines just mentioned in order to minimize bias and increase reliability. In this way, it attempts to arrive at 'true preferences' that can serve as intuition input coming close to what the famous philosopher John Rawls had in mind for his reflective equilibrium. But CREP does not end here.

3.6.3 Finding Coherence Between Public Preferences and Moral Theories

While we somehow promised that we would make the case for your opinion mattering, the view is not that you—or anyone else, for that matter—should be in the position to decide how moral dilemmas with self-driving cars should be regulated. That much we had to admit after the fruitful dialogue between Professor Principia and Professor Skeptico. Likewise, the results of well-designed studies offering reliable moral preferences should not determine the programming of self-driving cars on their own. That much Professor Skeptico had to admit to Professor Principia too. The next important step, he suggests, is to identify coherence between the laundered public preferences and moral principles as advanced by professional philosophers. In CREP, the relevant moral principles are the ones that are recommended by those ethical theories that are most endorsed in academic philosophy. It is the essence of what experts consider relevant (see also the overview in Fossa 2024, chap. 4.2).

This is not an unusual exercise, however, says Skeptico. Indeed, philosophers have been thinking about moral issues on all levels of abstraction for a very long time—for centuries, if not for millennia. Just think of Aristotle, for example, who lived in the fourth-century BC. Modern-day philosophy still finds inspiration in his writings on what makes people and their character traits good, or virtuous. Striving purposefully for certain ideals, such as excellence or dedication to the common good, makes for a good and, thus, moral life, according to him. Skeptico knows that there is even a New Virtue Ethics advocated by modern philosophers that builds on Aristotelian ideas (MacIntyre 1981; Hursthouse 1999). However, like Nyholm (2018), Professor Principia admits that virtue ethics does not seem to contribute much to the ethics of self-driving cars. This does not mean that the history of ethics is irrelevant, however, replies *Skeptico*: Other ethical traditions are very relevant, for instance, utilitarianism (Bentham, Mill) and deontology (Kant). Do you remember our discussion in Chapter 2, dear Principia? Virtue ethics, utilitarianism, and deontology are arguably the three families of ethical thinking that have emerged from this traditional philosophical debate as the most serious candidates for the best or most appropriate moral theory; and there are many attempts to combine (e.g., Parfit 2011) or refine them (e.g., Scanlon's contractualism in Chapter 2).

Listen to our colleagues, says Skeptico: Finding coherence between laundered public preferences and moral principles means the following:

> The idea is that the more a given public preference coheres with more theories and principles, the greater the justification and legitimacy it will have. Why it should have greater legitimacy should be clear: a policy that can be justified via multiple frameworks [...] can be endorsed by, and justified to, a greater number of people.
>
> *(Savulescu, Gyngell, and Kahane 2020, 659)*

So after laundering moral preferences found in empirical research, the task is to inquire whether the 'clean' preferences are in line (or in contrast) with the most reasonable moral theories, asks Principia. That is correct, confirms Skeptico, CREP works like this: If something is neither supported by the most reasonable moral theories nor reflected by public preferences: Reject it. If it is supported by preferences but not by theory, that is better, but you should still reject it. If it is supported by moral theory but contrary to public preferences, that is better than nothing but you should still reject it. If it is supported by theory but not supported by preferences (because we have no data about them), it might be ok, but ideally preferences should be researched. And finally, if theories and preferences are in favor: Bingo! For anything in between (a few theories/preferences in some studies but not in all in favor and a few against), then both options tend to be allowed. So have I got it right, asks Principia. Generally speaking, support for a policy is greater the more it reflects public preferences and the more it coheres with moral theories, correct? Yes, you got it, confirms Skeptico. It also works the other way around: The less a policy reflects public preferences and the less it is supported by moral theories, the weaker the case for it.

So do you see, dear Principia, your worry that people actually have discriminatory preferences and that paying attention to what the people say might lead to discriminatory policies is combatted by the combination of laundering the input and searching for coherence between preferences and moral theory in CREP? As you well know, none of the reasonable moral theories would support a policy to save members of one ethnic group but not another, or to discriminate against people on the basis of their faith, disability, or social status. Since CREP helps us dismiss policies that do not cohere with the most reasonable moral theories regardless of how widespread they are, such discriminatory policies would be off the table, elaborates Skeptico.

In Chapter 2, we discussed utilitarianism, deontology, and contractualism in the context of the permissibility of trade-offs between human lives. This is an important issue for which Bentham and Kant have well-grounded arguments for contradictory positions—always saving the many vs. never allowing trade-offs between human lives. However, this dialogue between Professor Principia and Professor Skeptico should not leave you with the impression that utilitarianism and deontology are contradictory moral theories. In fact, although the kind of moral thinking suggested by these theories differs fundamentally, it is not the case that they imply conflicting solutions for many moral problems. In a similar fashion, Principia and Skeptico also share much common ground. Nevertheless, as we saw in Chapter 2, there are important dilemma situations with self-driving cars where Principia and the specific theory she argues for on the one hand and Skeptico arguing for another theory on the other would each suggest programming different solutions. In such a case, it would be very interesting to see what the moral preferences found in empirical studies suggest. This might be the factor that decides which outcome of two well-established moral theories should be followed in the programming of self-driving cars.

Another important aspect is that different philosophers would likely arrive at different judgments about some particularly difficult cases, not only our philosopher friends Principia and Skeptico. Not so much because they rely on different moral theories, but because the interpretation, specification, and application of moral theories in the attempt to solve particular moral problems is not a mechanical process at all. It requires a lot of expertise and a certain sense of appropriateness (Paulo 2016). In Chapter 2, precisely this kind of philosophical work was made explicit by the discussion of particularly difficult issues between Professor Principia and Professor Skeptico.

Following the CREP method, support for a public policy is greater the more it reflects the public preferences of laypeople (found in sound empirical studies) and the more it coheres with moral theories (advanced by experts). The less a policy reflects public preferences and the less it is supported by moral theories, the weaker the case for it. Policies should be dismissed that neither reflect public preferences nor cohere with the most reasonable moral theories, as should policies that reflect public preferences but do not cohere with the most reasonable moral theories. This leaves open a wide range of possibilities, namely those that have some (but no strong) support from public preferences and cohere with some moral theories (but not with others). For these situations, CREP does not 'have a ready recipe to offer' (Savulescu, Gyngell, and Kahane 2020, 659).

Is this a shortcoming of the CREP method? It is probably fairer to say that this concession merely illustrates the boundaries of the possible contribution of ethics to questions of public policy. But that does not mean that such questions cannot be decided. In addition to the CREP method, we can use a particularly important area of the law as a tie-breaker.

3.6.4 *Human Rights as a Tiebreaker*

We began this chapter with Professor Principia's concern that the law has been neglected in the empirical study of moral preferences. It is not just famous cases involving moral decisions about life and death that are important in this respect. Human rights should also play a role when discussing how to regulate the programming of moral dilemmas in such situations (Kirchmair 2023). Indeed, questions of the life and death of 'innocent road users' should not be discussed in isolation from the law. More specifically, human rights should be included in the equation: Moral preferences should not be drawn solely from empirical studies and moral theories, as the CREP method suggests. Human rights should be added as a contributing factor. Why?

The law is neither irrelevant nor the only decisive argument in resolving moral dilemmas. Nevertheless, fundamental legal norms such as human rights have been put in place through long and arduous social processes. Human rights, thus, represent a kind of long-established international agreement on fundamental values. While it may be fair to say that human rights in general, and some specific human rights in particular, are not universal as they are not found all over the world, this is also true for public preferences. They seem to be different, at least to a certain extent, depending on whereabouts in the world we ask laypeople what they think the right thing to do is. Thus, when we speak of public preferences that are presumably representative of the whole world, we should also refer to universally recognized human rights. If we focus on public preferences found only in studies with a regional focus, regional human rights should serve as a reference point.

It is true that moral preferences do not necessarily have to coincide with the law, but we should be concerned when they violate human rights. Remember, we were even concerned when they were somehow influenced by minor biases. It may not be automatically amoral, or even immoral, to violate the law, but the chances are high that what violates human rights is also immoral. The least we can expect is to be given reasons why some preferences should be taken seriously as moral preferences even though they violate human rights. Simply stating that they would be moral because they were laundered moral preferences found in studies is not convincing in such a case. While we noted at the beginning of this chapter that the law can change and therefore does not stand in the way of morality per se, human rights do not change so easily and so frequently. Moreover, human rights represent the most fundamental values of those societies where they are respected and they are typically the foundation upon which other norms are built, if they emerge, for instance, in the same breath as a constitutional system of a society is shaped. Human rights are not the same as (measuring) public preferences or moral theorizing, true, but they are sufficiently similar that it seems justified to make them count in the CREP methodology as well, at least as a third element in the equation.

To this end, we propose extending the CREP method to include human rights as a tie-breaker. When policies are evaluated from a moral point of view, we check whether the public preferences found in empirical studies are consistent with moral theories. But if a certain policy is supported only by some public preferences and is consistent with some, but not all, moral theories, then human rights can be the decisive factor: If the policy in question violates human rights, then it should be rejected. Human rights are therefore decisive in the event of unclear outcomes produced by CREP.

3.7 Using the CREP Method for Our Purposes

So what are the laundered public preferences and how well do they cohere with moral theories?

In this short book, we are of course not able to provide a comprehensive overview (let alone a balanced analysis) of all empirical studies that are potentially relevant to the questions addressed in the book. We will reap the fruits of the discussions between Professor Principia and Professor Skeptico but nevertheless limit our discussion to some of the most recent studies that we deem most reliable and informative. Of course, new studies can reveal new findings that would make it necessary to update the assessment we present here. We have to live with that—and so do you. This provisional nature should not lead anybody to believe that empirical findings are irrelevant per se. That would be a severe misjudgment. What else should you base your assessment on if not on the best currently available studies? This is science, and the whole scientific enterprise works like that. We consider to be true what corresponds to currently available knowledge—like the earth being round. The fact that we simply cannot foresee future knowledge does not render current knowledge meaningless.

A similar warning is warranted for the part on moral theory. We do not know what the correct moral theory is (if there is one). But we do know which ones are considered the best candidates by those who have spent their whole careers dealing with this question. These are the theories we are going to look at. But as we indicated in Chapter 2, it is not always clear what a particular moral theory says about a specific moral problem. Theories need to be interpreted, specified, and applied, and at each of these levels, there are different opinions among proponents of the theory. Our task is not to resolve these disagreements but to identify where they arrive at the same recommendations with regard to moral dilemmas with self-driving cars, and where they do not.

3.7.1 Social Status

Public Preferences

The large-scale Moral Machine Experiment found that there is a preference to save those with a higher social status (e.g., a doctor) instead of people with a lower social status (e.g., a homeless person). When the car can either save a doctor or a homeless person, a majority of people prefer programming that leads to the doctor being saved and the homeless person being killed. This is a quite strong preference the researchers find worldwide (Awad et al. 2018).

However, the study also revealed cultural differences. These can be described as occurring across three distinct clusters, the 'Western cluster' (North America and many European countries), the 'Eastern cluster' (many countries in the Far East), and the 'Southern cluster' (Latin America and countries with a French influence). The most important differences across these clusters concern the weight that subjects attributed to certain moral preferences. Regarding social status, the study suggested that subjects from the Eastern cluster attribute less weight to sparing high-status people than subjects from the other clusters, although it is still significant in the Eastern cluster. So this preference might not be universally strong. However, the Moral Machine Experiment found that it is strong in the Western and Southern clusters.

Bigman and Gray (2020) replicated the main finding with a similar study design in which the participants (all from the Western cluster) could only choose between either saving the person with a low social status or the one with a high social status. In this forced-choice design, they found that 79.9% would rather save a person with a high social status.

However, when Bigman and Gray (2020) gave test persons a third option to answer, namely to treat all potential victims equally, they found that people prefer this equality option (84.7%) over the option to save those with the higher social status (12.7%). Other studies (De Freitas and Cikara 2021; De Freitas et al. 2020) found that fewer than 20% of the participants prefer self-driving cars to decide based on any social category such as race, gender, status, or age. Instead, they prefer to let the cars decide randomly.[8]

Thus, the preference to save those with higher social status is only found in studies with a forced-choice design. When offered the possibility to treat all potential victims equally, participants prefer this option. Here it proved important to launder the empirical findings. Forced-choice designs have limitations that make it problematic for their results to inform CREP. Hence, the empirical findings with a sound methodology suggest not to differentiate between people with different social status.

Moral Theories

As mentioned in Chapter 2, even Bonnefon, one of the architects of the experiment, said that the category of social status was meant to show the 'limits of an exercise such as Moral Machine for informing public policy.' For the researchers, '[c]itizens' preferences are not always acceptable from an ethical point of view, and the purpose of Moral Machine is therefore not to replace ethics committees' (Bonnefon 2021, 115). So, apparently, the researchers did regard it as obvious that social status should not be a basis for differential treatment in the programming of self-driving cars. Their intention to include it as a category in their experiment in the first place was, arguably, to highlight the limits of their approach.

Based on the small number of studies on this question, we can say Bonnefon's premise seems to be false. Citizens do not seem to have a preference to save those with a higher social status over those with a lower status, at least not if they are allowed to treat them equally. But what about social status from a moral point of view?

According to utilitarianism, the only thing that matters morally is the consequences of an action, how much pleasure it produces, and how much pain it avoids. The question is which action results in the greatest happiness for the greatest number. The social status of potential victims does not seem to make a difference in terms of utility.[9] So, for utilitarians, social status should not be a basis for differential treatment.

With its emphasis on human dignity, Kantian deontology demands that all humans be treated equally. It clearly prohibits any distinction based on personal features such as social status, age, ethnicity, etc. Similarly, Scanlon's contractualism prohibits any such distinction. There is no possible moral principle that would allow a preference for people with a higher social status over those with a lower status, because all people with a lower social status would have strong reasons for rejecting this principle. And indeed, their reasons for doing so would outweigh anyone's reasons for accepting the principle. That is why such principles are inappropriate according to contractualism.

Let us consider another version of a contract theory, namely Rawls' theory. Somewhat simplified, Rawls' view is that the morally just action is the one we would choose if we did not know who we would be in the situation under consideration. The idea is that we imagine to decide what to do from behind the so-called 'veil of ignorance' (Rawls 2005; Freeman 2019). Veiled from any knowledge about our own social status, we would certainly want no difference to be made between people based on their social status.

Summing up, all moral theories yield the same verdict: From a moral point of view, no difference should be made between people based on their social status. This verdict is in line with the public preference found in (laundered) empirical studies (without a forced-choice design).

3.7.2 Gender

Public Preferences

The Moral Machine Experiment found that there is a preference to save women rather than men. When the car can either save a man or a woman, a majority of people prefer programming that leads to the woman being saved and the man killed. This is the seventh strongest preference the researchers found worldwide (Awad et al. 2018)—and they took this one more seriously than their finding about social status. Regarding cultural differences, this preference to save women rather than men is higher in the Southern cluster than it is in the Western and Eastern clusters, but it is still significant in these two as well.

Bigman and Gray (2020) also replicated this finding with a study design in which the participants (all from the Western cluster) could only choose between either saving a man or a woman. In this forced-choice design, they found that 87.7% would rather save the woman. Similarly, Sütfeld et al. (2017) discovered that 80% of their test participants would rather sacrifice men in direct comparisons between men and women.

However, studies that departed from the forced-choice design did not reveal the same preference. Bigman and Gray (2020) found that people prefer the equality option (92.6%) over the option to save women (6.4%). Similarly, De Freitas and Cikara (2021) and De Freitas et al. (2020) found that fewer than 20% of their participants prefer self-driving cars to decide based on any social category (such as race, gender, status, or age). Instead, they prefer to let the cars decide randomly.[10]

Studies that used a forced-choice design found a clear preference to save women rather than men. But studies that offered a third answer option found evidence for a public preference to prefer neither women nor men.

Moral Theories

Utilitarianism, Kantianism, and the two versions of contractualism support this finding. From the point of view of utilitarianism, the gender of potential victims does not seem to make a difference in terms of utility. So, for utilitarians, gender should not be a basis for differential treatment. Kantian deontology, too, demands that all humans be treated equally and prohibits any distinction based on personal features such as social status, gender, age, ethnicity, etc. Similarly, Scanlon's contractualism prohibits any such distinction because all men (or women) would have strong reasons for rejecting any principle that would recommend differential treatment, and their reasons would outweigh anyone's reasons for accepting the principle. Finally, in Rawls' theory, viewed from behind the 'veil of ignorance,' not knowing which gender one has would result in not wanting a difference to be made between people based on their gender.

Summing up, all moral theories yield the same verdict: From a moral point of view, no difference should be made between people based on their gender. This verdict is in line with the laundered public preference found in empirical studies, at least in those without a forced-choice design.

3.7.3 Age

Public Preferences

The Moral Machine Experiment revealed that there is a preference to save younger people rather than the elderly. When the car can either save a child or an elderly person, the majority of people prefer programming that leads to the young person being saved and the elderly person being killed. This is the third strongest preference worldwide (Awad et al. 2018). Regarding cultural differences, the study found that participants from the Eastern cluster attributed less weight to sparing younger people rather than older ones than participants from the other two clusters. So this preference might not be universally strong, although it is still significant in the Eastern cluster. In contrast, the Moral Machine Experiment discovered that it is very strong in the Western and Southern clusters.

Bigman and Gray (2020) replicated this main finding with a study design in which the participants could only choose between either saving a child or an elderly person. In this forced-choice design, 96.1% would rather save the child. Faulhaber et al. (2019) found a similar preference to save the young.

However, things become more uncertain when we consider studies that do not use a forced-choice design. Bigman and Gray (2020) found that people prefer the equality option (61.1%) over the option to save the child (38.8%). De Freitas and Cikara (2021) and De Freitas et al. (2020) found that fewer than 20% of their participants prefer self-driving cars to decide based on any social category (such as race, gender, status, or age). Instead, they prefer to let the cars decide randomly. Note, however, that these studies did not report findings about the different categories mentioned.[11]

Thus, studies that used a forced-choice design found a clear preference to save the young rather than the elderly. Among the few studies that offered a third answer option, the majority of people show a preference not to prioritize the young over the old (although 40% are still in favor of prioritizing the young despite the possibility of treating the young and the old equally). All in all, there is also evidence for a public preference not to decide on the basis of age.

Moral Theories

Kantian deontology with its emphasis on human dignity demands that all humans be treated equally. It clearly prohibits any distinction based on personal features such as social status, age, ethnicity, etc. Similarly, Scanlon's contractualism prohibits any such distinction. There is no possible moral principle that would allow younger people to be preferred over the elderly because all elderly people would have strong reasons for rejecting this principle. Their reasons for doing so would outweigh the reasons of the young for accepting the principle.

But is this fair? As discussed in Chapter 2 (Section 2.5), everyone belongs to a certain age group, just as they belong to a certain religious (or non-religious) group, have a certain skin tone, and a certain gender. The difference to other forms of group-based discrimination, however, is that we all start young and grow older (if we are lucky). In contrast, we do not usually start with one religion, skin tone, or gender and then change them every decade or so (although some do).

This thought is captured by utilitarianism with its emphasis on the consequences of an action, how much pleasure it produces, and how much pain it avoids. The question is which action results in the greatest happiness for the greatest number. Utilitarians would prioritize younger people, because, on average, younger lives lead to more utility in the future. So, for utilitarians, age should be a basis for differential treatment.

What about Rawlsian theory? It is not clear what an individual would want when judged from the 'veil of ignorance.' On the one hand, they would not know if they were young or old. That individual would certainly not want to be the person that gets killed. The chances of both options being equal speak for random choice or similar impartial decision rules. On the other hand, who would mind less about being killed by a car—a young person or an older one? The thought that the young person would have had less life, and hopefully more expected life to look forward to, speaks for saving the young (Savulescu, Gyngell, and Kahane 2020, 661; see also Leben 2018, chap. 5).

Summing up, Kantian deontology and Scanlon's contractualism suggest that, from a moral point of view, no difference should be made between people based on their age. Rawls' theory can be understood to support this view. However, it can also be understood to support the utilitarian judgment that age should be a basis for differential treatment. The moral verdict is thus unclear. This conclusion mirrors the mixed empirical findings on public preferences.

As the CREP method does not yield a clear result, we turn to human rights law. The human right to equality, and particularly the obligation of non-discrimination, stands in the way of treating children and elderly people differently. The principle of equality includes a ban on age discrimination in most jurisdictions.[12] This human right includes the right not to be discriminated against on the basis of age in life-and-death decisions.

3.7.4 Numbers

Public Preferences

The Moral Machine Experiment revealed that there is a preference to save as many people as possible. When the car can either save one person or a group of two or more, a large majority of the participants prefer programming that leads to the group of two or more being saved and the one person being killed. This is the second strongest preference the researchers found worldwide, the strongest preference being to save humans rather than animals (Awad et al. 2018). There were no significant cultural differences regarding this preference to save as many people as possible.

Bigman and Gray (2020) replicated this main finding with a study design in which the participants could only choose between either saving one person or a group. In this forced-choice design, 99.6% would rather save the many. Faulhaber et al. (2019) found a similar preference (more than 90%). Huang, Greene, and Bazerman (2019) also discovered that more than 80% of participants prefer to save the many (even when asked to decide from behind the 'veil of ignorance').

When Bigman and Gray (2020) gave participants a third option, namely to treat all potential victims equally, they found that only 17.9% preferred this equality option; a majority (81.6%) still preferred to save as many lives as possible. Similarly, Bodenschatz, Uhl, and Walkowitz (2021) ascertained that 67.6% would rather save the many than opt for a random choice between the two options (30.6%).

Studies that used a forced-choice design found a clear preference to save as many lives as possible. Studies that offered a third answer option found a somewhat lower but still strong preference to save the many.

Moral Theories

As discussed in Chapter 2, utilitarianism supports the public preference to save the many. Which action results in the greatest happiness for the greatest number? If every person's life has the same (or roughly similar) levels of well-being, then the saving of two lives leads to more utility than the saving of one, likewise three more than two, and so on. Utilitarians would therefore save the greater number.

Things are more complicated in Kantian deontology. Here is one of Kant's formulations of the categorical imperative: 'So act that you use the humanity, whether in your own person or in the person of any other, always at the same time as an end, never merely as a means' (Kant 1996, 80). The German Ethics Commission has the Kantian categorical imperative in mind when it says in the context of self-driving cars that 'the sacrifice of innocent people in favour of other potential victims is impermissible, because the innocent parties would be degraded to mere instrument and deprived of the quality as a subject' (Ethics Commission 2017, 18). According to this position, human lives must not be offset against each other. The idea of human dignity blocks all attempts to engage in trade-offs between human lives.

The implications for self-driving cars are clear: They must not be programmed to always save the greatest number.

However, other Kant scholars argue that deontology can be interpreted in such a way that the one person is not 'used' in order to save the larger group because the car's killing of the one person does not causally contribute to the saving of the larger group. The car would also swerve so that the group does not get hit if the one person were not standing there (Kleingeld 2020; Guyer 2014; Kerstein 2013). So these Kantian deontologists come to the conclusion that it is permissible to program self-driving cars to always save the greatest number.

What does Rawlsian theory say? Behind the 'veil of ignorance,' an individual simply cannot know whether they are part of the larger group or the person standing alone. They would thus opt for programming that leads to the greatest chance of surviving; the probability of being part of a group is higher than being the one person so that they would opt to save the greater number.[13]

What about Scanlon's contractualism? Remember that contractualism suggests asking whether a principle can be justified in free agreement with other agents. It is essential to consider the objections that other agents would raise against our principles and take them seriously. In Chapter 2, we mentioned a case in which a rescuer can either save the life of one individual who is trapped on one rock or the lives of five individuals who are trapped on a second rock but cannot save the lives of all six. This Rock case is very similar to the Bystander version of the Trolley cases. Whereas utilitarianism would recommend saving a greater number of individuals, contractualism points out that each of the five individuals on the second rock can only appeal to their own personal claim of wanting to be saved. Since the individual on the first rock can equally well appeal to their personal claim of wanting to be saved, contractualism seems to yield the result that it is permissible for the rescuer to save either the many or the few. What does this mean for programming self-driving cars? It would suggest letting the cars decide randomly, or preventing them from swerving, resulting in the death of those who happen to be in front of the car, be it the one person or the group.

Within contractualism, people disagree on whether this solution is correct or not. Scanlon himself argues that it is false. If we only had to choose between two individuals on two different rocks, it would indeed be correct to consider both alternatives permissible. But if we considered both alternatives permissible in this particular case, the five people could complain that their case was being treated as though there were only one person on their rock, with four of them being treated as though they did not matter at all. In other words, in scenarios like the Rock case, the number functions as a tie-breaker. Scanlon would thus favor programming that leads to the greatest number of people being saved.

Other contractualists disagree. They maintain, for example, that the tie-breaker argument only works when an individual's claim to not being killed is considered together with the claim of another member of the group. Like utilitarianism, contractualism attributes a key role to the well-being of individuals. However, in contrast to utilitarianism, it does so in a decidedly non-aggregative way. According to contractualism, moral deliberation is based on taking the perspective of different individuals in turn and considering their objections. In this process, only individuals and individual advantages and disadvantages are discussed. Crucially, no individual can invoke the potential overall loss of a group in their own defense. This individualist restriction seems to be violated in Scanlon's application of his own theory to the numbers problem (Otsuka 2006).

Recently, Henning (2024) developed a new contractualist argument for Scanlon's position to save the largest number of people. He points out that people do not only have an interest in surviving or not getting hurt. Another morally relevant interest is their having a say in the processes that lead to decisions that affect them. In many dilemma situations, individuals' interests to survive or not to get hurt compete. One can only save person A or person B but not both. Individuals' procedural claims do not compete in the same way. Everyone can have an equal voice in the process, even if they do not like the outcome. It is, thus, not necessary to adjudicate among these procedural interests. A winning party does not need to be selected which then has sole influence over the decision-making process, while none of the others has a say at all. On this basis, Henning suggests relying on the decision procedure of majority rule: 'If each affected person votes for the option in which she herself is saved, and if we let majority rule determine the option we realize, then we will realize an option that saves at least as many people as any other option' (Henning 2024, 759). Now the problem is that we cannot ask people whether they want to be saved in actual cases. This is why Henning simply assumes that everyone would want to be saved. In the context of self-driving cars, we have a better option: We can draw on the empirical findings about public preferences. Of course, these are not preferences stated or revealed in actual dilemma situations but reactions to fictional scenarios. But we think it is fair to say that these empirical findings provide the best available basis for any guess about what people would want in dilemma situations. As summarized above, studies with and without a forced-choice design revealed a strong public preference for saving as many lives as possible. This supports Henning's assumption and his conclusion that as many lives as possible should be saved.

This solution gets further support from the discussion of the Trolley problem in Chapter 2. There we argued that, despite the long-standing debate on how to solve the Trolley problem, the whole debate is premised on the view that certain answers to paradigmatic Trolley cases are correct. For example, it is assumed to be permissible for the bystander to hit the switch in the Bystander case. Approx. 90% of laypeople and approx. 63% of professional philosophers find it permissible to hit the switch. Now take the case of a self-driving car that will run into a group of five (killing them all) if it is programmed to stay on track or will kill one of them if it is programmed to swerve. Assuming that this scenario is morally similar to the Bystander case, programming self-driving cars to let the car swerve (killing one) in order to save a larger group of people is also permissible. If the two cases are morally similar, they warrant the same moral judgment. The discussion of the Trolley problem, thus, also supports the view that as many lives as possible should be saved (Paulo 2023).

Summing up, Kantian deontology as understood by the German Ethics Commission and one interpretation of contractualism both speak against programming self-driving cars in such a way that would always lead to saving as many lives as possible. They would suggest that the cars stay on track, killing whoever is in front of them, or that the decision is made randomly. All of the other moral theories discussed here—other interpretations of Kantian deontology and contractualism, utilitarianism, *Rawls*' theory, and the discussion of the Trolley problem—speak for programming the car to save as many lives as possible. All in all, the majority of the (interpretations of) moral theories discussed support the public preference found in studies with and without a forced-choice design, namely to save as many lives as possible. Arguably, human rights do not stand in the way of such a conclusion (Kirchmair 2023). However, this judgment is not shared by all legal scholars. According to most legal scholars in Germany, for instance, the fundamental right to human dignity is interpreted in line with the abovementioned reasoning of the German Ethics Commission, namely that it prohibits the weighting of human lives and thus the prioritization of a larger group over a smaller one.

3.8 Conclusion

In Chapter 3, we tried to keep the promise we made to you at the beginning of our book. In the first few lines, we tried to focus your attention on moral dilemmas. We even dared to say that what you think about them should count and be of interest when resolving such dilemmas. However, we did not promise easy choices, nor did we say that solving moral dilemmas is easy. So you should make up your mind before you decide. While we made you aware of some of the main arguments in the debate in Chapter 2, in Chapter 3 we wanted to show you how what you think matters. It is not only what you think but also what your neighbor thinks that matters to us. More generally, in Chapter 3 we wanted to show you the importance of what laypeople think about morality, and how such lay moral preferences can be reconciled and balanced with the views of experts—engineers, ethicists, and lawyers.

We began and ended with references to the law, emphasizing that it should not distract us too much if the law is not on the same page as moral preferences and moral experts. Nevertheless, we also introduced human rights as a tiebreaker for situations in which the CREP method does not yield clear results.

However, the main message of this chapter is that morality is not something that can be found only in the opinions of laypeople or in the opinions of experts. The most appropriate understanding of morality for our purposes is to try and combine both elements, your opinions and the opinions of experts. Morality is not to be found in the 'ivory tower' of ethical theorizing; nor is it simply to be found in what (the majority of) laypeople think is right. Even if important parts of our discussion are controversial among philosophers, there are practical and theoretical reasons for combining lay and expert opinions. Using the method of reflective equilibrium introduced by one of the most famous philosophers of the last century, John Rawls, we showed you how to combine empirical evidence about 'what people think' with what moral philosophers think.

A group of ethicists recently proposed a sensible method for taking account of both public morality and traditional moral theory: 'Collective Reflective Equilibrium in Practice (CREP).' This method restricts findings about public moral preferences to empirical studies that are well-designed and meet high methodological standards. This means that no decisions made in a state of agitation or excitement or biased by cognitive traps should count. Only methodologically sound preferences are valid. The preferences ascertained in this way are combined with the moral principles from the most important moral theories. Following the CREP method, we can conclude that the more a policy—in our case the programming of self-driving cars for accidents—reflects the public preferences of laypeople (found in sound empirical studies) and the more it is consistent with moral theories (advanced by experts), the greater the support for the policy. The less a policy reflects public preferences and the less it is supported by moral theories, the weaker the case for it. Policies should be rejected that neither reflect public preferences nor are consistent with the most reasonable moral theories, as should policies that reflect public preferences but are not consistent with the most reasonable moral theories.

In our case, we showed that, from a moral point of view, no distinction should be made between people on the basis of their social status or gender in relation to accidents involving self-driving cars. If you are wondering why it is necessary to emphasize that, you might want to take a look at Chapter 2 where we discussed that there were many studies actually testing whether such characteristics do make a difference—and some, indeed, found that these were the case. However, the judgment we just offered to you is in line with the public preference found in (laundered) empirical studies (without a forced-choice design).

When it came to age, however, the picture was less clear. While Kantian deontology and Scanlon's contractualism suggest that, from a moral point of view, no distinction should be made between people on the basis of their age, Rawls's theory can be understood as supporting both options, i.e., saving children at the expense of the elderly or treating them equally. The moral verdict from the perspective of experts is, therefore, unclear. In fact, this conclusion reflects the mixed empirical evidence on public preferences. However, when human rights and the principle of non-discrimination are taken into account, the conclusion is that children should not be saved at the expense of the elderly.

If we are solely discussing programming that differentiates between group sizes, then only Kantian deontology, as understood by the German Ethics Commission, and an interpretation of contractualism argue against programming that always leads to saving as many lives as possible. All of the other moral theories discussed here—other interpretations of Kantian deontology and contractualism, utilitarianism, Rawls's theory, and the discussion of the trolley problem—argue for programming that always leads to saving as many lives as possible. Overall, the majority of (interpretations of) moral theories discussed support the public preference found in studies with and without a forced-choice design, namely to save as many lives as possible. Human rights do not stand in the way of such an interpretation either. In short, here you can see that your opinion (if you belong to the vast majority in this case) should count. It may not be much, you might think, but it is not nothing either. Given the large number of car accidents, such a finding has a very large impact. And so policymakers and their regulations should pay close attention to this finding. This is what we will look at in the final Chapter 4.

Notes

1 While there are many critical perspectives on the Moral Machine Experiment, Principia's line of reasoning follows loosely the criticism articulated by the philosopher John Harris (Harris 2020).
2 Skeptico's line of reasoning loosely follows our response to Harris (see Paulo, Möck, and Kirchmair 2023).
3 There is a long-standing debate about the conceptual separation of law and morality that basically aims at shielding the (process of) law(-making) from external influences (see, e.g., the classical essay by Hart, 1958; for an overview, Ratnapala, 2017, Chapter 8). Our argument reverses that point: The (process of) morality(-making) should be shielded from external influences such as the law.
4 For details about Miller's methodology, see Pölzler (2023) and Schramme (2018).
5 Note that what we describe in the text is only the "narrow" version of reflective equilibrium; there is also a more complex "wide" version of the method. For these notions and for an overview of the recent debate about reflective equilibrium, see Tersman (2018) and Daniels (2016).
6 This is the ideal situation. In the early 2010s, the so-called "replication crisis" rocked psychological research (and later almost any field that relies on empirical studies), because the vast majority of studies published in reputable journals could not be replicated by other researchers. This casted doubt on the reliability of such research. However, the crisis has also led to a significant

improvement in the methods used and the transparency of the process—some have even spoken of a 'credibility revolution' (see, e.g., Korbmacher et al. 2023).

7 For a more detailed discussion of reliability, see Paulo (2020a).

8 In studies that we conducted together with Yochanan Bigman, we found that more than 80% prefer a programming option that treats people of different social status impartially. Impartial treatment means that certain considerations—here, a person's social status—do not play any role in making the decision (Gert 1995; Jollimore 2022). Because this study has not yet been peer reviewed or published, one should not yet rely on it.

9 In Chapter 2, we discussed the possibility that participants might have understood social status in terms of how beneficial someone is for society. For example, a doctor (high social status) is likely to be more beneficial for society than a criminal (low social status), and being more beneficial might well correlate with higher utility. However, there is no strong or even necessary connection between social status and how beneficial a person is for society. This is why social status itself does not link to utility.

10 In the yet unpublished studies that we conducted together with Yochanan Bigman, we found that more than 90% prefer a programming option that treats men and women impartially.

11 In the yet unpublished studies that we conducted together with Yochanan Bigman, we found that approx. 35% prefer a programming option that treats the child and the elderly person impartially. When this impartiality option is combined with the decision-rule to remain inactive (i.e., not to swerve), this combination is preferred by approx. 52%.

12 Human rights of older persons are, however, not a universal standard. See, for instance, the statement of the UN independent expert on the enjoyment of all human rights by older persons, Claudia Mahler, who says that

> older persons remain largely invisible in international human rights law due to the lack of recognition of ageism and of an explicit prohibition of age-based discrimination. Among United Nations treaties, only two of the core human rights treaties contain explicit references to age.
> (Mahler 2023)

13 For a more detailed discussion of what the Rawlsian theory suggests for the ethics of self-driving cars, see Leben (2017).

References

Abarbanell, Linda, and Marc D. Hauser. 2010. "Mayan Morality: An Exploration of Permissible Harms." *Cognition* 115 (2): 207–24. https://doi.org/10.1016/j.cognition.2009.12.007.

Ahlenius, Henrik, and Torbjörn Tännsjö. 2012. "Chinese and Westerners Respond Differently to the Trolley Dilemmas." *Journal of Cognition and Culture* 12 (3–4): 195–201.

Arras, John D. 2007. "The Way We Reason Now: Reflective Equilibrium in Bioethics." In *The Oxford Handbook of Bioethics*, edited by Bonnie Steinbock, 46–71. Oxford; New York: Oxford University Press.

Awad, Edmond, Sohan Dsouza, Richard Kim, Jonathan Schulz, Joseph Henrich, Azim Shariff, Jean-François Bonnefon, and Iyad Rahwan. 2018. "The Moral Machine Experiment." *Nature* 563 (7729): 59. https://doi.org/10.1038/s41586-018-0637-6.

Banerjee, Konika, Bryce Huebner, and Marc Hauser. 2010. "Intuitive Moral Judgments Are Robust across Variation in Gender, Education, Politics and Religion: A Large-Scale Web-Based Study." *Journal of Cognition and Culture* 10 (3–4): 253–81. https://doi.org/10.1163/156853710X531186.

Bartels, Daniel M., and David A. Pizarro. 2011. "The Mismeasure of Morals: Antisocial Personality Traits Predict Utilitarian Responses to Moral Dilemmas." *Cognition* 121 (1): 154–61. https://doi.org/10.1016/j.cognition.2011.05.010.

Bigman, Yochanan E., and Kurt Gray. 2020. "Life and Death Decisions of Autonomous Vehicles." *Nature* 579 (7797): E1–2. https://doi.org/10.1038/s41586-020-1987-4.

Bodenschatz, Anja, Matthias Uhl, and Gari Walkowitz. 2021. "Autonomous Systems in Ethical Dilemmas: Attitudes toward Randomization." *Computers in Human Behavior Reports* 4 (August): 100145. https://doi.org/10.1016/j.chbr.2021.100145.

Bonnefon, Jean-François. 2021. *The Car That Knew Too Much: Can a Machine Be Moral?* Cambridge, MA: MIT Press.

Bostyn, Dries, Sybren Sevenhant, and Arne Roets. 2018. "Of Mice, Men, and Trolleys: Hypothetical Judgment Versus Real-Life Behavior in Trolley-Style Moral Dilemmas – Dries H. Bostyn, Sybren Sevenhant, Arne Roets, 2018." *Psychological Science* 29 (7): 1084–93. https://doi.org/10.1177/0956797617752640.

Cikara, Mina, Rachel A. Farnsworth, Lasana T. Harris, and Susan T. Fiske. 2010. "On the Wrong Side of the Trolley Track: Neural Correlates of Relative Social Valuation." *Social Cognitive and Affective Neuroscience* 5 (4): 404–13. https://doi.org/10.1093/scan/nsq011.

Costa, Albert, Alice Foucart, Sayuri Hayakawa, Melina Aparici, Jose Apesteguia, Joy Heafner, and Boaz Keysar. 2014. "Your Morals Depend on Language." *PLOS ONE* 9 (4): e94842. https://doi.org/10.1371/journal.pone.0094842.

Daniels, Norman. 2016. "Reflective Equilibrium." In *The Stanford Encyclopedia of Philosophy*, edited by Edward N. Zalta, Summer 2020. Metaphysics Research Lab, Stanford University. https://plato.stanford.edu/archives/sum2020/entries/reflective-equilibrium/.

De Freitas, Julian, and Mina Cikara. 2021. "Deliberately Prejudiced Self-Driving Vehicles Elicit the Most Outrage." *Cognition* 208 (March): 104555. https://doi.org/10.1016/j.cognition.2020.104555.

De Freitas, Julian, Sam E. Anthony, Andrea Censi, and George A. Alvarez. 2020. "Doubting Driverless Dilemmas." *Perspectives on Psychological Science* 15 (5): 1284–88. https://doi.org/10.1177/1745691620922201.

Dreier, Horst. 2007. "Grenzen des Tötungsverbotes – Teil 1." *JuristenZeitung* 62 (6): 261–70. https://doi.org/10.1628/002268807780282749.

Ethics Commission. 2017. "Report on Automated and Connected Driving." June 1, 2017. www.bmvi.de/SharedDocs/EN/publications/report-ethics-commission.pdf?__blob=publicationFile.

Faulhaber, Anja K., Anke Dittmer, Felix Blind, Maximilian A. Wächter, Silja Timm, Leon R. Sütfeld, Achim Stephan, Gordon Pipa, and Peter König. 2019. "Human Decisions in Moral Dilemmas Are Largely Described by Utilitarianism: Virtual Car Driving Study Provides Guidelines for Autonomous Driving Vehicles." *Science and Engineering Ethics* 25 (2): 399–418. https://doi.org/10.1007/s11948-018-0020-x.

Fossa, Fabio. 2024. *Ethics of Driving Automation: Artificial Agency and Human Values*. 1st ed. 2023 Edition. Cham: Springer.

Francis, Kathryn B., Sylvia Terbeck, Raluca A. Briazu, Agi Haines, Michaela Gummerum, Giorgio Ganis, and Ian S. Howard. 2017. "Simulating Moral Actions: An Investigation of Personal Force in Virtual Moral Dilemmas." *Scientific Reports* 7 (1): 1–11. https://doi.org/10.1038/s41598-017-13909-9.

Freeman, Samuel. 2019. "Original Position." In *The Stanford Encyclopedia of Philosophy*, edited by Edward N. Zalta, Summer 2019. Metaphysics Research Lab, Stanford University. https://plato.stanford.edu/archives/sum2019/entries/original-position/.

Friesdorf, Rebecca, Paul Conway, and Bertram Gawronski. 2015. "Gender Differences in Responses to Moral Dilemmas: A Process Dissociation Analysis." *Personality and Social Psychology Bulletin* 41 (5): 696–713. https://doi.org/10.1177/0146167215575731.

Fumagalli, Manuela, Maurizio Vergari, Patrizio Pasqualetti, Sara Marceglia, Francesca Mameli, Roberta Ferrucci, Simona Mrakic-Sposta, et al. 2010. "Brain Switches Utilitarian Behavior: Does Gender Make the Difference?" *PLOS ONE* 5 (1): e8865. https://doi.org/10.1371/journal.pone.0008865.

Gert, Bernard. 1995. "Moral Impartiality." *Midwest Studies in Philosophy* 20 (1): 102–28. https://doi.org/10.1111/j.1475-4975.1995.tb00307.x.

Gleichgerrcht, Ezequiel, and Liane Young. 2013. "Low Levels of Empathic Concern Predict Utilitarian Moral Judgment." *PLOS ONE* 8 (4): e60418. https://doi.org/10.1371/journal.pone.0060418.

Gold, Natalie, Andrew Colman, and Briony Pulford. 2014. "Cultural Differences in Responses to Real-Life and Hypothetical Trolley Problems." *Judgment and Decision Making* 9 (1): 65–76.

Guyer, Paul. 2014. *Kant*. 2nd ed. London: Routledge.

Harris, John. 2020. "The Immoral Machine." *Cambridge Quarterly of Healthcare Ethics* 29 (1): 71–79. https://doi.org/10.1017/S096318011900080X.

Hart, Herbert L. A. 1958. "Positivism and the Separation of Law and Morals." *Harvard Law Review* 71 (4): 593–629. https://doi.org/10.2307/1338225.

Hauser, Marc, Fiery Cushman, Liane Young, R. Kang-Xing Jin, and John Mikhail. 2007. "A Dissociation Between Moral Judgments and Justifications." *Mind & Language* 22 (1): 1–21. https://doi.org/10.1111/j.1468-0017.2006.00297.x.

Henning, Tim. 2024. "Numbers without Aggregation." *Noûs* 58 (3). https://doi.org/10.1111/nous.12475.

Henrich, Joseph. 2020. *The Weirdest People in the World: How the West Became Psychologically Peculiar and Particularly Prosperous*. New York: Farrar Strauss & Giroux.

Hilgendorf, Eric. 2018. "The Dilemma of Autonomous Driving: Reflections on the Moral and Legal Treatment of Automatic Collision Avoidance Systems." In *Digitization and the Law*, edited by Eric Hilgendorf and Jochen Feldle, 57–90. Baden-Baden: Nomos.

Hoffrage, Ulrich. 2022. "Overconfidence." In *Cognitive Illusions*, edited by Rüdiger Pohl, 3rd ed., 235–54. London: Routledge.

Horvath, Joachim, and Steffen Koch. 2020. "Experimental Philosophy and the Method of Cases." *Philosophy Compass* 16 (1): e12716. https://doi.org/10.1111/phc3.12716.

Huang, Karen, Joshua D. Greene, and Max Bazerman. 2019. "Veil-of-Ignorance Reasoning Favors the Greater Good." *Proceedings of the National Academy of Sciences*, November. https://doi.org/10.1073/pnas.1910125116.

Huemer, Michael. 2008. "Revisionary Intuitionism." *Social Philosophy and Policy* 25 (1): 368–92. https://doi.org/10.1017/S026505250808014X.

Hursthouse, Rosalind. 1999. *On Virtue Ethics*. Oxford; New York: Oxford University Press.

Jollimore, Troy. 2022. "Impartiality." In The Stanford Encyclopedia of Philosophy, edited by Edward N. Zalta, Summer 2022. Metaphysics Research Lab, Stanford University. https://plato.stanford.edu/archives/sum2022/entries/impartiality/.

Kahane, Guy, Jim A. C. Everett, Brian D. Earp, Miguel Farias, and Julian Savulescu. 2015. "'Utilitarian' Judgments in Sacrificial Moral Dilemmas Do Not Reflect Impartial Concern for the Greater Good." *Cognition* 134: 193–209.

Kant, Immanuel. 1996. "Groundwork of the Metaphysics of Morals (1785)." In *Practical Philosophy*, edited and translated by Mary J. Gregor, 37–108. The Cambridge Edition of the Works of Immanuel Kant. Cambridge: Cambridge University Press. https://doi.org/10.1017/CBO9780511813306.007.

Kerstein, Samuel J. 2013. *How to Treat Persons*. Oxford: Oxford University Press.

Kirchmair, Lando. 2023. "How to Regulate Moral Dilemmas Involving Self-Driving Cars: The 2021 German Act on Autonomous Driving, the Trolley Problem, and the Search for a Role Model." *German Law Journal* 24 (7): 1184–208. https://doi.org/10.1017/glj.2023.83.

Kleingeld, Pauline. 2020. "A Kantian Solution to the Trolley Problem." In *Oxford Studies in Normative Ethics*, edited by Mark Timmons, 204–28. Oxford: Oxford University Press. 10.1093/oso/9780198867944.003.0010.

Korbmacher, Max, Flavio Azevedo, Charlotte R. Pennington, Helena Hartmann, Madeleine Pownall, Kathleen Schmidt, Mahmoud Elsherif, et al. 2023. "The Replication Crisis Has Led to Positive Structural, Procedural, and Community Changes." *Communications Psychology* 1 (1): 1–13. https://doi.org/10.1038/s44271-023-00003-2.

Lanteri, Alessandro, Chiara Chelini, and Salvatore Rizzello. 2008. "An Experimental Investigation of Emotions and Reasoning in the Trolley Problem." *Journal of Business Ethics* 83 (4): 789–804. https://doi.org/10.1007/s10551-008-9665-8.

Leben, Derek. 2017. "A Rawlsian Algorithm for Autonomous Vehicles." *Ethics and Information Technology* 19 (2): 107–15. https://doi.org/10.1007/s10676-017-9419-3.

———. 2018. *Ethics for Robots: How to Design a Moral Algorithm*. New York: Routledge.

Liao, S. Matthew, Alex Wiegmann, Joshua Alexander, and Gerard Vong. 2012. "Putting the Trolley in Order: Experimental Philosophy and the Loop Case." *Philosophical Psychology* 25 (5): 661–71. https://doi.org/10.1080/09515089.2011.627536.

Machery, Edouard. 2017. *Philosophy Within Its Proper Bounds*. Oxford; New York: Oxford University Press.

MacIntyre, Alasdair C. 1981. *After Virtue: A Study in Moral Theory*. Notre Dame: University of Notre Dame Press.

Mahler, Claudia. 2023. "Make UDHR Promises a Reality for Older Persons: UN Expert." www.ohchr.org/en/statements/2023/09/make-udhr-promises-reality-older-persons-un-expert.

Miller, David. 1999. *Principles of Social Justice*. Cambridge, MA: Harvard University Press.

———. 2020. "Needs-Based Justice. Theory and Evidence." In *Empirical Research and Normative Theory: Transdisciplinary Perspectives on Two Methodical Traditions between Separation and Interdependence*, edited by Max Bauer and Malte Meyerhuber, 273–94. Berlin: De Gruyter.

Nyholm, Sven. 2018. "The Ethics of Crashes with Self-Driving Cars: A Roadmap, I." *Philosophy Compass* 13 (7): e12507. https://doi.org/10.1111/phc3.12507.

Otsuka, Michael. 2006. "Saving Lives, Moral Theory, and the Claims of Individuals." *Philosophy and Public Affairs* 34 (2): 109–35. https://doi.org/10.1111/j.1088-4963.2006.00058.x.

Parfit, Derek. 2011. *On What Matters*. Vol. 1. 3 vols. Oxford; New York: Oxford University Press.

Paulo, Norbert. 2016. *The Confluence of Philosophy and Law in Applied Ethics*. Basingstoke: Palgrave Macmillan.

———. 2020a. "Moral Intuitions Between Higher-Order Evidence and Wishful Thinking." In *Higher-Order Evidence and Moral Epistemology*, edited by Michael Klenk, 54–77. London: Routledge.

———. 2020b. "The Unreliable Intuitions Objection Against Reflective Equilibrium." *Journal of Ethics*, 24: 333–53. https://doi.org/10.1007/s10892-020-09322-6.

———. 2023. "The Trolley Problem in the Ethics of Autonomous Vehicles." *The Philosophical Quarterly* 73 (4): 1046–66. https://doi.org/10.1093/pq/pqad051.

Paulo, Norbert, Leonie Möck, and Lando Kirchmair. 2023. "The Use and Abuse of Moral Preferences in the Ethics of Self-Driving Cars." In *Experiments in Moral and Political Philosophy*, edited by Hugo Viciana, Antonio Gaitán, and Fernando Aguiar, 290–309. London: Routledge.

Petrinovich, Lewis, and Patricia O'Neill. 1996. "Influence of Wording and Framing Effects on Moral Intuitions." *Ethology and Sociobiology* 17 (3): 145–71. https://doi.org/10.1016/0162-3095(96)00041-6.

Petrinovich, Lewis, Patricia O'Neill, and Matthew Jorgensen. 1993. "An Empirical Study of Moral Intuitions: Toward an Evolutionary Ethics." *Journal of Personality and Social Psychology* 64 (3): 467–78. https://doi.org/10.1037/0022-3514.64.3.467.

Pölzler, Thomas. 2023. "How to Do Empirical Political Philosophy: A Case Study of Miller's Argument for Needs-Based Justice." *Erkenntnis*, November. https://doi.org/10.1007/s10670-023-00747-7.

Ratnapala, Suri. 2017. *Jurisprudence*. 3rd ed. Cambridge: Cambridge University Press.

Rawls, John. 1951. "Outline of a Decision Procedure for Ethics." *Philosophical Review* 60 (2): 177–97.

———. 2005. *A Theory of Justice*. Original Edition. Cambridge, MA: Belknap Press.

Santoni de Sio, Filippo. 2017. "Killing by Autonomous Vehicles and the Legal Doctrine of Necessity." *Ethical Theory and Moral Practice* 20 (2): 411–29. https://doi.org/10.1007/s10677-017-9780-7.

Savulescu, Julian, Christopher Gyngell, and Guy Kahane. 2020. "Collective Reflective Equilibrium in Practice (CREP) and Controversial Novel Technologies." *Bioethics*. https://doi.org/10.1111/bioe.12869.

Scanlon, Thomas M. 2002. "Rawls on Justification." In *The Cambridge Companion to Rawls*, edited by Samuel Freeman, 139–67. Cambridge; New York: Cambridge University Press.

Schramme, Thomas. 2018. "Mill and Miller: Some Thoughts on the Methodology of Political Theory." *CONTEMPORARY POLITICAL THEORY* 17 (1): 85–90.

Schwitzgebel, Eric, and Fiery Cushman. 2015. "Philosophers' Biased Judgments Persist Despite Training, Expertise and Reflection." *Cognition* 141: 127–37.

Strohminger, Nina, Richard L. Lewis, and David E. Meyer. 2011. "Divergent Effects of Different Positive Emotions on Moral Judgment." *Cognition* 119 (2): 295–300. https://doi.org/10.1016/j.cognition.2010.12.012.

Sütfeld, Leon R., Richard Gast, Peter König, and Gordon Pipa. 2017. "Using Virtual Reality to Assess Ethical Decisions in Road Traffic Scenarios: Applicability of Value-of-Life-Based Models and Influences of Time Pressure." *Frontiers in Behavioral Neuroscience* 11. https://doi.org/10.3389/fnbeh.2017.00122.

Svenson, Ola. 1981. "Are We All Less Risky and More Skillful than Our Fellow Drivers?" *Acta Psychologica* 47 (2): 143–48. https://doi.org/10.1016/0001-6918(81)90005-6.

Swann, William B., Ángel Gómez, John F. Dovidio, Sonia Hart, and Jolanda Jetten. 2010. "Dying and Killing for One's Group: Identity Fusion Moderates Responses to Intergroup Versions of the Trolley Problem." *Psychological Science* 21 (8): 1176–83. https://doi.org/10.1177/0956797610376656.

Swift, Adam, and Stuart White. 2008. "Political Theory, Social Science and Real Politics." In *Political Theory: Methods and Approaches*, edited by David Leopold and Marc Stears, 49–69. Oxford: Oxford University Press.

Tersman, Folke. 2018. "Recent Work on Reflective Equilibrium and Method in Ethics." Philosophy Compass. https://doi.org/10.1111/phc3.12493.

Tversky, Amos, and Daniel Kahneman. 1974. "Judgment under Uncertainty: Heuristics and Biases." *Science* 185 (4157): 1124–31. https://doi.org/10.1126/science.185.4157.1124.

Valdesolo, Piercarlo, and David DeSteno. 2006. "Manipulations of Emotional Context Shape Moral Judgment." *Psychological Science* 17 (6): 476–77.

Wiegmann, Alex, Yasmina Okan, and Jonas Nagel. 2012. "Order Effects in Moral Judgment." *Philosophical Psychology* 25 (6): 813–36. https://doi.org/10.1080/09515089.2011.631995.

Wolff, Jonathan. 2019. *Ethics and Public Policy: A Philosophical Inquiry*. 2nd ed. New York: Routledge.

Zamzow, Jennifer L., and Shaun Nichols. 2009. "Variations in Ethical Intuitions." *Philosophical Issues* 19 (1): 368–88. https://doi.org/10.1111/j.1533-6077.2009.00164.x.

4

OUR TIPS FOR POLICYMAKERS

4.1 Introduction

The aim of this book is to prepare readers to take positions on difficult issues, not spontaneously or in the heat of the moment, but after considering the relevant arguments. In Chapters 1 to 3, we limited ourselves to enabling our readers to make a well-informed decision for themselves. This is important not only for laypeople but even more so for those who make policy in this area as they have to deal with some of the most difficult moral issues facing societies worldwide in the near future, namely moral dilemmas involving self-driving cars. In this chapter, we provide insights for policymakers around the world on what *we* believe is important when regulating moral dilemmas involving self-driving cars.

We summarize some of the main findings of the book in the form of recommendations for the regulation of self-driving cars. That is, we briefly recapitulate the problem and the main arguments for the most promising solutions for resolving moral dilemmas involving self-driving cars through regulation. We do not promise that policymakers will find it easy to decide what is right and what is wrong after reading these tips. But they will know which positions are morally defensible (and which are not). They will also understand why it is morally problematic not to take a decision, and why ethicists, lawyers, and—most importantly—lawmakers need to know what people think about moral dilemmas involving self-driving cars.

4.2 Regulation Is Necessary and Urgent

Self-driving cars should be allowed on public roads: All in all, it seems desirable to allow self-driving cars on the roads. *Functioning* self-driving cars can increase mobility for those who cannot drive, and they can improve traffic control, which could contribute to more environmentally friendly mobility. Most importantly for the subject of this book, self-driving cars will significantly reduce the number of road casualties. Because of this promise, it is important to regulate self-driving cars in a way that does not deter the general public. If some decisions relating to self-driving cars in fatal accidents violate

DOI: 10.4324/9781003485216-4

important moral preferences that people hold, they may be outraged enough to oppose the use of such cars. The moral reasons for having self-driving cars can be explained by the old moral precept *primum non nocere* ('first, do no harm'). From a legal perspective, the right to life (one of the most fundamental individual rights) entails an obligation of the state to take precautionary measures to prevent dangerous situations. Allowing self-driving cars on public roads reduces harm; therefore, they should be allowed on public roads.

Car accidents will still happen, and moral dilemmas are inevitable: Despite the promise of improved safety, car accidents will happen. As we learned from historical examples, even if cars only travel at 4 mph, there will be fatalities. Given the technical dangers and the problems of mixed traffic, the only way to reduce the risk of accidents to zero would be to stop car traffic altogether. This will not happen. The innovation of self-driving cars gives us a new, amazing, and, at the same time, difficult opportunity to decide in advance what should happen in which accident situation. So we have to find a regulation for a very uncomfortable situation. There is no silver bullet; there is no way that the regulation of programming will be favorable to all those potentially affected by it. Against this background, regulation faces a difficult challenge. It has to decide how to distribute the negative costs of self-driving cars, and they cannot be distributed evenly. For the first time in history, we as a society can decide in advance who to harm and who to save. It is about making decisions on moral dilemmas after careful reflection.

A (new) regulation is necessary: The benefits of the new technology are a step forward, but they require new regulations. Self-driving car technology brings with it innovations that challenge the way traffic regulation works. While people used to be the addressees of traffic laws, it is now the machines. It is not that self-driving cars will be self-managing autonomous agents, but regulating the programming of self-driving cars is still different from regulating human traffic.

Individuals should not decide for themselves how to program their self-driving cars for dilemma situations: While people generally prefer self-driving cars that promote the 'public good,' in the sense that they are programmed to save as many people as possible, they would rather not buy a self-driving car that is programmed to sacrifice its passengers for the greater good. This 'social dilemma' can be pre-empted by legal regulation: A fairly balanced distribution of the negative burdens of self-driving cars should be mandatory for every self-driving car allowed on public roads. If people were allowed to program their cars themselves, the coordination advantages of these cars might also disappear. The ability to coordinate the behavior of self-driving cars is an important factor in their superior safety.

It is not an option not to decide: Self-driving cars will have to be programmed one way or another. They will inevitably harm some and save others. The rise of self-driving cars is therefore both a gift and a burden. Self-driving cars offer the chance to save many lives. But they also force us to decide who must die in order to save others in dilemma situations. This difficulty cannot be overcome by leaving such moral dilemmas unregulated. No decision is also a decision because accidents will still happen, and any programming will lead to negative results. More importantly, the ability to control self-driving cars and to program them to achieve certain results means that car manufacturers can use this power to favor their interests and those of their customers, which is unlikely to be in the interest of society as a whole. Pedestrians and cyclists, in particular, could lose out because car manufacturers are likely to be motivated to produce cars that favor their passengers since they are the ones buying the cars. The question of how to make decisions about moral dilemmas involving self-driving cars must, therefore, be answered by the law.

The establishment of regulation is an urgent matter: Establishing appropriate regulation is very important and urgent. The technology of self-driving cars is developing rapidly, and the power to decide who lives and who dies will become an urgent matter. The issue is highly topical because its central question will very soon affect all of us directly.

4.3 Regulation Needs to Talk to Code

Regulations for machines must be designed differently than regulations for humans: For example, while humans can read between the lines, understand abstract language, and consider abstract principles as a fallback solution when there is no concrete rule, machines cannot. Rules for machines need to be written in a language that machines can understand, in very precise binary language, anticipating any situation, and providing concrete regulation for these. Therefore, the regulation of self-driving cars must be comprehensive and speak to code (Sütfeld, Bronson, and Kirchmair 2025).

Regulations for machines must strike the balance between defining the limits for car manufacturers and avoiding too many restrictions on competition: Despite the challenge and need for concrete regulation, it would be a mistake to prescribe overly specific programming for self-driving cars by law. Competition between different car manufacturers can help to find the best software solution for different situations. However, the regulation should not be too abstract. Otherwise, there is too much scope for private companies to prioritize their own interests. The regulation needs to talk to the code (instead of to people). This requires a concrete regulation. Instead of an abstract language, self-driving cars need a binary language, thus abandoning abstract regulation.

Several issues need to be considered simultaneously: Regulation must take place, time, action, and responsibility into account as well as various options and probabilities. This is a lot and requires complex balancing operations. Such operations need to be specified in the regulation because how these issues are balanced will, to a large extent, determine the outcome.

Probabilities matter: Beyond these issues, there are other things to consider when regulating the accident behavior of self-driving cars. We will have to deal with probabilities. All other things being equal, we need to balance two different outcomes along the lines of the calculated probability that they will materialize. So it might be better to prioritize the action that has a lower probability of a negative outcome than the alternative. A 5% chance of a negative outcome is better than a 90% chance of a negative outcome.

The balance between different aspects must be regulated: This means that, overall, we need a regulation that requires self-driving cars to consider all of the aspects and recommendations, etc. mentioned here and, importantly, also provides guidelines on how to balance conflicting aspects. The regulation must prescribe how different probabilities and different severities of injury, or rather likelihoods of such injury, are to be balanced so as to minimize risk.

4.4 Recommendations

The following recommendations draw on the analysis we provided in Chapters 1 to 3 in particular, as well as on our work on self-driving cars in general. There we introduced a specific method to fruitfully combine the insights from available studies on public moral preferences and relevant moral theories. The method is called Collective Reflective Equilibrium in Practice (CREP, see Savulescu, Gyngell, und Kahane (2021)).

Rule #1: Reduce risk: Unfortunately, self-driving cars will not be able to completely eliminate fatal accidents. What is important, however, is to minimize risks. This aim should serve as an overarching principle. This means that, whenever possible, self-driving cars will avoid dangerous situations by slowing down, taking evasive action, or choosing a different route. Reducing risk is an imperative of common sense, a general principle of constitutional democracies, and a requirement of the right to life. Accordingly, all moral theories, as well as the moral preferences of laypeople, underline this. In fact, all current laws on the protection of human life contain some provision to this effect. Both the German High-Level Ethics Commission and the Expert Group on Specific Ethical Issues Raised by Driverless Mobility, which advises the European Commission, recommend that self-driving cars should be programmed to reduce risk.

Rule #2: Prioritize people: In conflicts between different goods, it is important to give priority to people. The protection of humans is regarded as more important than the protection of things or animals. Again, this can generally be deduced from any modern legal system as well as from empirical studies and moral theories. Both ethics commissions (EU and German) and the current laws also suggest this prioritization.

Rule #3: Social status and gender should not be taken into account: In a moral dilemma, most personal characteristics should be set aside. In Chapter 3, we demonstrated that there is widespread support from moral theories as well as from public preferences for this view regarding the social status and the gender of potential victims. These personal characteristics should not be the deciding factor in prioritizing a person's protection. The same applies to other typical grounds of discrimination such as ethnicity, religion, etc. This view is shared by the German ethics commissions and the current legal regulations. The commission advising the European Commission is largely silent on such issues. For age as a personal characteristic, see rule #4.

Rule #4: Age should not be taken into account: Age might be seen as an exception to the general rule not to take personal characteristics into account (rule #3). In Chapter 3, we demonstrated that the empirical evidence shows a public preference for treating the young and the elderly equally. At the same time, studies show that age is judged somewhat differently from other classic grounds of discrimination. Studies on age show a significantly weaker preference for equal treatment or impartiality between potential victims. In those cases where there is a preference for equality, it is much less pronounced than for the other grounds of discrimination. In addition, studies with a forced-choice design indicate a rather clear preference for favoring the young. It is also noteworthy that no study shows a preference for the elderly. With regard to age, the relevant moral theories come to different conclusions. Some prohibit any differentiation on the basis of age. Other theories consider age—in contrast to other personal characteristics such as gender or social status—to be a morally relevant ground for treating potential victims differently. They argue for a preferential treatment of the young. No moral theory advocates favoring the elderly. This mirrors the somewhat mixed empirical findings for public preferences. The CREP method used to integrate public preferences and moral theories thus issues no particularly clear recommendation about whether or not it is permissible to take age into account when programming self-driving cars—in the sense of preferential treatment of the young, not the other way around. For such situations, we recommend extending the CREP method and using human rights as a tiebreaker. The human right to equality includes the right not to be discriminated against on the basis of age in life-and-death decisions. We thus recommend not allowing for differential treatment based on age.

Rule #5: Save as many lives as possible: As reported in Chapter 3, empirical studies show a strong public preference to save the many. When a car can either save one person or a group of two or more (at the cost of killing the former), people prefer programming that leads to the saving of as many people as possible. Most of the relevant moral theories support this public preference. Even those theories that speak against programming that always leads to the saving of as many lives as possible do not suggest saving the few. Rather, they suggest that the cars stay on track, killing whoever is in front of them, or that the decision is made randomly. Note that current legal regulations are silent on this issue, as is the commission advising the European Commission. The German ethics commission is divided on this point. For situations where the out-of-control car is heading toward a group and can only be redirected to hit a smaller number of people from that group (rather than the whole group), the Commission has suggested the possibility that programming to reduce the number of casualties would be morally acceptable. Our recommendation is that self-driving cars should be programmed to save as many lives as possible.

Rule #6: Respect rules: While no one advocates the death penalty for jaywalkers, some form of responsibility for one's actions is important. Everyone currently needs to comply with traffic laws, and this should not change with self-driving cars. Therefore, if someone causes an accident by their sudden, unexpected behavior such as running onto the street, the consequence should not be that someone else who has followed all of the rules has to pay, just because, say, this person is in a smaller group. This is important because this is how current traffic laws work. If this were to change, people would feel insecure about crossing the road even when the light is green because a larger group of jaywalkers could kill them. Another case illustrates why rules are important. Usually, traffic laws require those who ride a motorcycle to wear a helmet. So if a self-driving car has to decide whether to hit a motorcyclist wearing a helmet or a motorcyclist without a helmet, protecting your head with a helmet should not put you at risk. While the unhelmeted motorcyclist is certainly more vulnerable, such prioritization would discourage responsible behavior and be detrimental in the long run. In brief, the person who has caused mobility risks has to bear an extra burden. This should also be reflected in the programming rules for self-driving cars.

Rule #7: Reduce interference and avoid deliberate killing as much as possible: Most people are not in favor of conscious decisions to disadvantage, i.e., kill, someone. Therefore, it is preferable to follow 'fate' or, more technically, the status quo without interference (inaction as a conscious decision) rather than consciously kill someone.

4.5 Epilogue

If we had to pin down the most important finding of this book, perhaps it would be that most people have the moral preference that it is right to choose to protect the larger group over the smaller group. Some, but admittedly not all, moral theories also support such a solution. Interestingly, this is not what the current laws on self-driving cars like the German 2021 Act, say. We believe that this should be changed. The law should reflect the moral preferences of laypeople if they find some support in important moral theories and do not violate human rights. All available empirical studies seem to support such programming, even when—and this is important—participants in the studies are offered a third option and therefore makes such a study consistent with the current methodological state of the art, namely to treat both groups equally (despite the difference in size).

By contrast, we would hesitate to advise lawmakers around the world to favor younger over older lives, as the authors of the Moral Machine Experiment, for example, suggest, because this is controversial in moral theory, empirical evidence does not clearly support such a preference, and, finally, it violates human rights. While such a preference appears to be statistically significant in the Moral Machine Experiment and other studies, this changes radically when participants are given the option of treating younger and older lives equally.

In addition to these findings, and despite some empirical studies suggesting otherwise, most people do not want to differentiate between people with different personal characteristics. It would, therefore, be unwise to prioritize women over men, or doctors over the homeless.

Another important lesson is that self-driving cars need to be approached differently from humans. While humans can read between the lines and make meaningful interpretations based on general principles in rather specific life situations, computers still cannot. They work in binary code. To be concrete, it is important to keep in mind that the regulations for self-driving cars must be very specific. Vague and abstract rules, such as 'self-driving cars should minimize risks,' leave a lot of leeway to those programming the cars. But giving car manufacturers too much power is not the best solution either. They will take the interests of the company, and therefore the interests of their customers, much more into account than they should.

Three points—reducing harm (also to the extent that larger groups are more important than smaller groups), prioritizing human life, and prohibiting the saving of human life on the basis of personal characteristics—seem to be the bare minimum that should not be left to individuals or companies to decide. No matter how much you pay for your self-driving car, it does not buy you the right to program your car to save—or harm—a particular gender, skin color, or age. At least these elements of programming seem to be off the table for individual ethics.

But the most important finding, in our view, is that laypeople's opinions matter when it comes to moral dilemmas. This is particularly true in the context of moral dilemmas involving self-driving car accidents because such cars are likely to become part of our everyday lives. Therefore, all of us should have a say in the regulation of such sensitive issues as life-and-death decisions in which we may one day be involved.

The results of empirical studies may evolve with improved methodology or changing preferences over time and across regions. Moral theories are not static either. They depend on moral philosophers and their interpretations, which also happen to change over time. So your views are likely to evolve and adapt. If reading this book has contributed in some small way to a wider public debate on such an important issue—which enables taking your position, the position of your neighbor, and the opinion of experts into account—then we have achieved our goal.

References

Savulescu, Julian, Christopher Gyngell, and Guy Kahane. 2021. "Collective Reflective Equilibrium in Practice (CREP) and Controversial Novel Technologies." *Bioethics* 35 (7): 652–63. https://doi.org/10.1111/bioe.12869.

Sütfeld, Leon R., Joshua Bronson, and Lando Kirchmair. 2025. "Automated Vehicle Regulation Needs to Speak to Code, Not to Humans: Keeping Safety and Ethics in the Public Domain." Working Paper.

INDEX

Note: Endnotes are indicated by the page number followed by "n" and the note number e.g., 111n5 refers to note 5 on page 111.